SPSS Companion

for

Lattin, Carroll, and Green's

Analyzing Multivariate Data

Douglas L. Baker
James Lattin
Stanford University

THOMSON
BROOKS/COLE

Australia • Canada • Mexico • Singapore • Spain • United Kingdom • United States

Printed in the United States of America
1 2 3 4 5 6 7 08 07 06 05 04

Printer: Globus Printing

ISBN: 0-534-38226-6

For more information about our products, contact us at:
Thomson Learning Academic Resource Center
1-800-423-0563

For permission to use material from this text or product, submit a request online at **http://www.thomsonrights.com.**
Any additional questions about permissions can be submitted by email to **thomsonrights@thomson.com.**

Thomson Brooks/Cole
10 Davis Drive
Belmont, CA 94002-3098
USA

Asia
Thomson Learning
5 Shenton Way #01-01
UIC Building
Singapore 068808

Australia/New Zealand
Thomson Learning
102 Dodds Street
Southbank, Victoria 3006
Australia

Canada
Nelson
1120 Birchmount Road
Toronto, Ontario M1K 5G4
Canada

Europe/Middle East/South Africa
Thomson Learning
High Holborn House
50/51 Bedford Row
London WC1R 4LR
United Kingdom

Latin America
Thomson Learning
Seneca, 53
Colonia Polanco
11560 Mexico D.F.
Mexico

Spain/Portugal
Paraninfo
Calle/Magallanes, 25
28015 Madrid, Spain

Table of Contents

1 SPSS for Data Analysis

This SPSS Companion Guide demonstrates working examples for twelve multivariate analysis techniques discussed in the textbook Analyzing Multivariate Data by James Lattin, J. Douglas Carroll and Paul E. Green, published by Thomson Learning, 2003.[1] Companion Guide chapters match comparably numbered chapters within Analyzing Multivariate Data. Guide Chapter 2 therefore refers to discussion in the textbook Chapter 2 on matrix algebra, and so on. Readers primarily interested in substantive material on multivariate techniques should skip Chapter 2 and begin with Chapter 3 on regression analysis. Obtain Companion Guide data files from the CD-Rom accompanying Analyzing Multivariate Data.

SPSS techniques in this guide are demonstrated in a step-by-step fashion, accompanied by illustrations. Wherever possible techniques are demonstrated both as interactive menu choices and as text-edited syntax (SPSS programming language). Examples are taken from the SPSS 11.5 Base version. Users of previous Windows and MacIntosh SPSS versions should be able to adapt most examples.

SPSS Companion Guide examples refer to applications other than SPSS. Chapter 6 on confirmatory factor analysis and Chapter 10 on structural equation modeling employ the stand-alone student version of Amos. The current Amos student version can be obtained free from the Smallwaters Corporation homepage at http://www.smallwaters.com/amos/student.html. Chapter 7 on Multidimensional Scaling includes sample programs created by James Lattin for KYST, SINDSCAL and MDPREF. The KYST, SINDSCAL and MDPREF applications can be copied from the Analyzing Multivariate Data text CD-Rom.

The following conventions refer to file names and SPSS elements in this guide:

- Principal SPSS working environments are in lower case, as in 'Data Editor' and 'Syntax Editor'.
- SPSS menu and dialog box names are in bold, as in **File** menu. Menu choices are indicated by hyphens, as in the '**File-Save**' menu path. **File-Save** indicates the **File** menu is selected, and within it the **Save** menu item is selected.
- SPSS command keywords are in upper case, as in 'DESCRIPTIVES VARIABLES='.
- Data files, variable names, and user-specified command options are in lower case font, as in the 'Height' variable.

[1] SPSS Inc., 233 South Wacker Drive, Chicago, Ill 60606.

Example 1.1 *Managing SPSS Data files from the Data Editor window*

Copy women.sav from the Analyzing Multivariate Data CD-Rom and open. The Data Editor **Variable View** displays elements of the data dictionary in a spreadsheet-like format. The SPSS Data Editor window displays the working data file to be analyzed. Only one working data file can be open at any time.

women.sav - SPSS Data Editor

File Edit View Data Transform Analyze Graphs Utilities Window Help

	Name	Type	Width	Decimals	Label	Values	Missing	Columns
1	col1	Numeric	2	0		None	None	8
2	col2	Numeric	3	0		None	None	8
3								

Data View / Variable View

SPSS Processor is ready

Rename variables col1 and col2 as Height (Height in inches) and Weight (body weight in pounds) by directly text editing within the Name column. A descriptive label for each variable may be entered in the **Label** column. Variable name changes and variable labels are conveniences to make output easier to understand. They do not affect computation.

women.sav - SPSS Data Editor

File Edit View Data Transform Analyze Graphs Utilities Window Help

	Name	Type	Width	Decimals	Label	Values	Missing	Columns
1	height	Numeric	2	0		None	None	8
2	weight	Numeric	3	0		None	None	8
3								

Data View / Variable View

SPSS Processor is ready

Select the **Data View** tab at the bottom of window to display data values, also in a spreadsheet-like format of variables (columns) and cases (rows). Data values can be input or edited within the Data Editor.

women.sav - SPSS Data Editor

File Edit View Data Transform Analyze Graphs Utilities Window Help

1 : height 57

	height	weight	var	var	var	var	var	var	var
1	57	93							
2	58	110							
3	60	99							
4	59	111							

Data View / Variable View

SPSS Processor is ready

Click the **File-Save** menu to save data. The SPSS format (.sav extension) retains data values and the full data dictionary (variable names, labels missing value specifications). For example, saving the women.sav file retains the new variable names weight and height in place of col1 and col2.

Example 1.2 *Using SPSS Menus*

Most SPSS functions can be run from the Data Editor menus. The Data Editor **File**, **Edit**, **View**, **Window** and **Help** menus contain typical Windows menu features. Open data sets by following the **File-Open Data** menu path. To open spreadsheet files such as Excel (e.g., .xls), specify a target file extension and whether row #1 contains variable names. Open ASCII files through the **File-Read Text Data** menu path and a **Text Import** wizard. Most statistical analysis and graphing commands are generated from the **Analyze** and **Graphs** menus. The **Data**, **Transform** and **Utilities** menus contain SPSS data management and programming options.

Descriptive Statistics Command

Select the **Analyze-Descriptive Statistics** menu path to open the **Descriptives** dialog box. Designate target variables in the **Descriptives** dialog box by highlighting them. Then click the central box with the rightward-pointing arrowhead to designate items in the **Variables** list to be analyzed.

Click the **Options** button to open the **Descriptives: Options** dialog box, and choose which statistics to display.

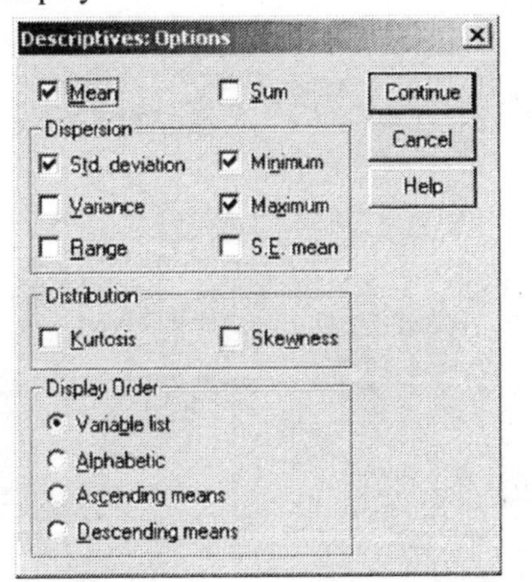

To obtain results click **Continue** and then **OK**.

Scatterplot Command

Follow the **Graphs-Scatter** menu path to generate the **Scatterplot** dialog box.

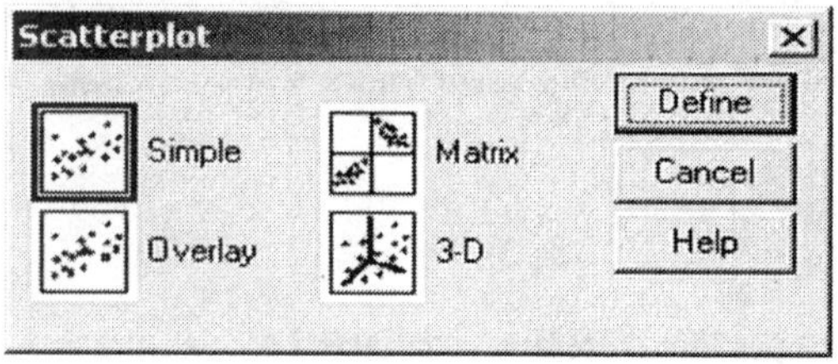

Click the **Simple** and **Define** buttons to display the **Simple: Scatterplot** dialog box and specify commands for a bivariate scatterplot. Again assign variables to the x and y axes by clicking boxes with directional arrows.

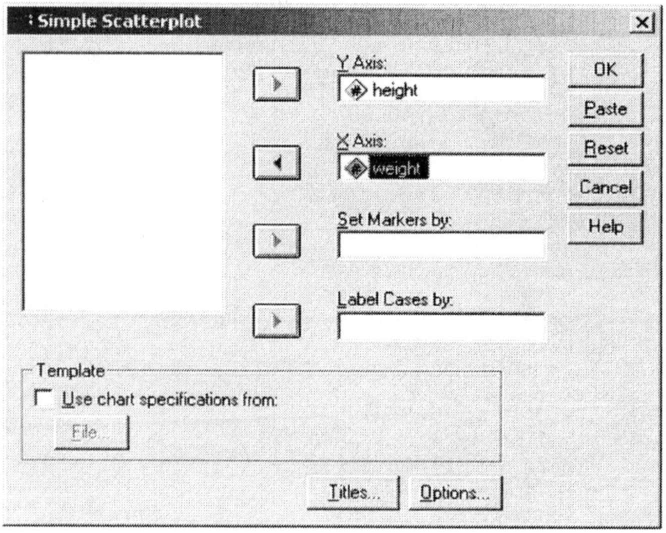

Click **Continue** and the **OK** to obtain results.

SPSS *Output*

The SPSS Output Viewer displays command results, including tables, graphs, warnings or error messages. The left hand side of the Viewer window contains an outline-ordered running log to allow convenient navigation of larger output runs. Output objects can be edited, copy-pasted or exported in a variety of formats such as html. Double click to edit, single-click to copy-paste or export.

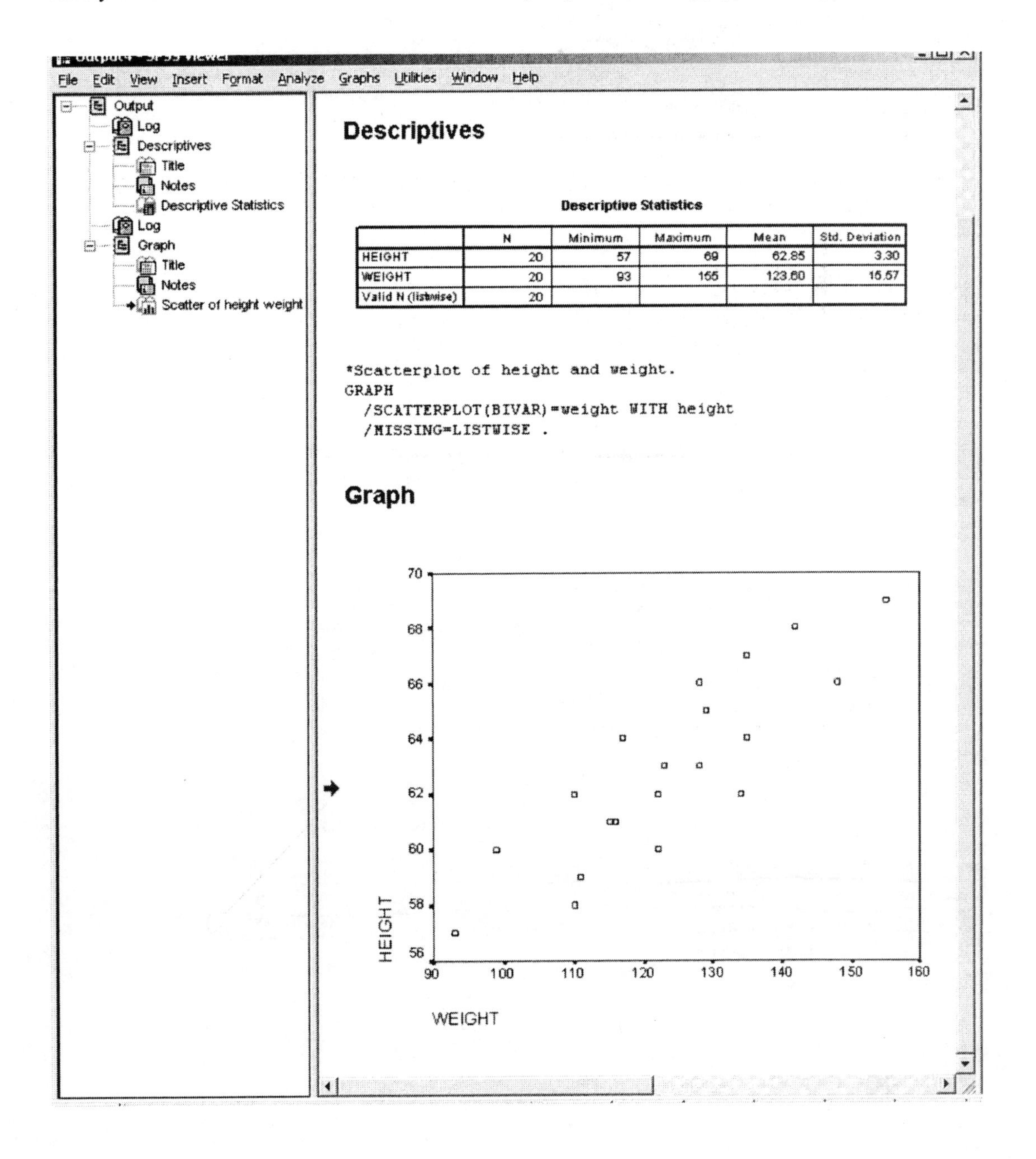

Descriptive Statistics

	N	Minimum	Maximum	Mean	Std. Deviation
HEIGHT	20	57	69	62.85	3.30
WEIGHT	20	93	155	123.60	15.57
Valid N (listwise)	20				

Example 1.3 *SPSS Syntax editing*

'Syntax' refers to the SPSS programming language and command elements. This grammatically-consistent set of rules controls every aspect of the SPSS application. The command is the basic syntax unit. A command is initiated by a keyword, such as DESCRIPTIVES, followed by subcommands, options, and completed by the command terminator, a period (.).

Generate default syntax with identical steps to Example 1.2, selecting the **Analyze-Descriptive Statistics** menu path to open the **Descriptives** dialog box. Designate target variables in the **Descriptives** dialog box. Click the **Paste** button to display the command set within an SPSS Syntax Editor window, where commands can be text-edited. Syntax can often be text edited in economical terms. For example SPSS accepts abbreviations for many keywords. The terse phrase 'des all' instructs SPSS to produce descriptive statistics for all variables in the working data set. 'Des all' means exactly the same thing syntactically, and will produce exactly the same output, as the menu-pasted command:
'DESCRIPTIVES VARIABLES=height weight /STATISTICS=MEAN STDDEV MIN MAX .'

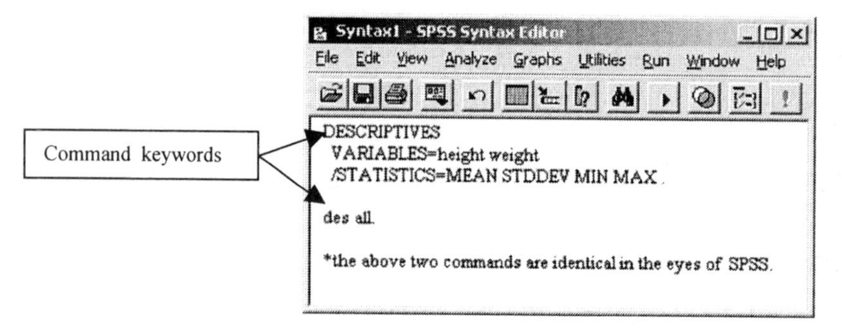

It is usually helpful to document important programming steps. Any text within the syntax file preceded by an asterisk is considered by SPSS to be a comment and ignored in processing. In concept the comment line is a separate command to ignore text following the keyword '*' until terminated, like all SPSS commands, by a period (.).

SPSS commands conform to a relatively small number of general rules. Those most relevant to completing the exercises in Analyzing Multivariate Data include the following:

- SPSS syntax is not case sensitive.

- A command must begin with the Command name keyword (for example DESCRIPTIVES). One cannot run subcommands independent of the command name.
- Similarly, subcommands begin with a keyword (for example VARIABLES).
- Subcommands may typically be in any order.
- Command expressions may be separated by spaces, and lines can be broken at any allowable space.
- Each command must end with a terminator—a period (.) is the default terminator. SPSS is sometimes willing to accept the last non-blank character it encounters before a subsequent recognizable command.
- Subcommands have two delimiters:
    - The '=' sign identifies specifications which follow the subcommand keyword
    - The '/' slash distinguishes one subcommand from another.
- Multiple specifications within a subcommand typically have blank spaces as delimiters.
- Command line maximum length is 80 characters.
- Most commands used with the Analyzing Multivariate Data text problems are immediately read through the dataset, line-by-line, in the order they are executed.
- A command that modifies or creates new variables must precede a command that analyzes those modifications.

Running syntax

To execute (run) SPSS syntax first highlight the commands that you want to run in the Syntax Editor window. Then click the Run button (right-pointing triangle) on the syntax window toolbar. Or click on the Run menu and select a choice.

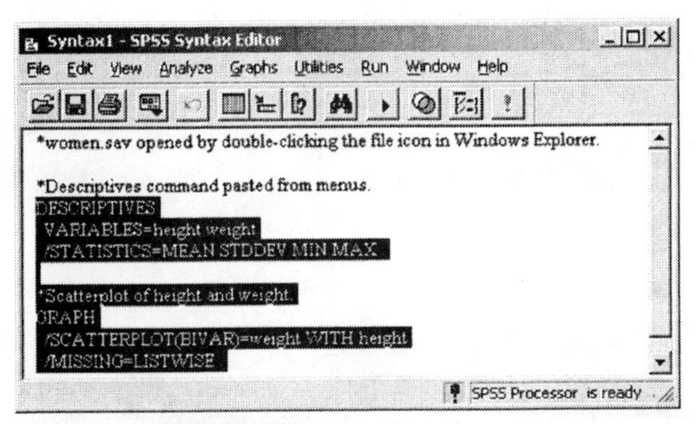

Executing an SPSS command causes the working data file to be read and command instructions carried out. In the example above first the DESCRIPTIVES command is executed, followed by the GRAPH command.

Save syntax files with the **File-Save** menu path.

Debugging syntax

Warning and error messages can be useful clues to identify mistakes such as misspellings. Warnings sometimes accompany a syntax that runs but has logical flaws. At other times SPSS issues a warning but refuses to read the data file as in the output below, where the variable Height was misspelled 'heght.' SPSS cannot recognize the variable and refuses to run the Descriptives command.

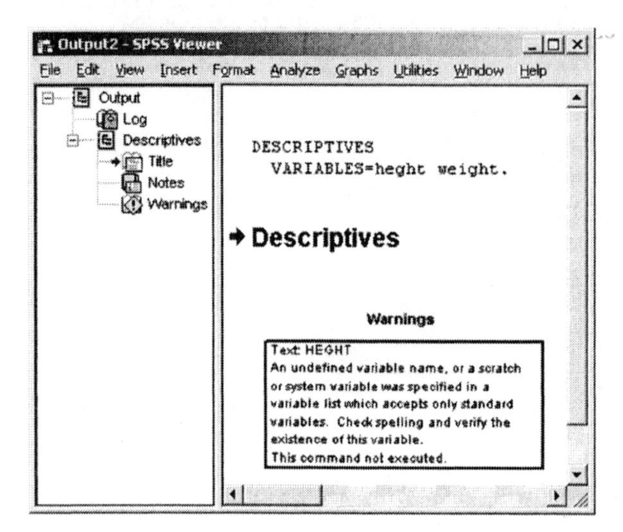

Revise and rerun syntax command-by-command to locate the origin of errors and confirm they have been resolved.

spss.jnl - Notepad

File Edit Format Help

```
*Examine frequencies of the variables height and weight.
FREQUENCIES VARIABLES = heght weight.
Text: HEGHT
*>An undefined variable name, or a scratch or system variable was specified
*>in a variable list which accepts only standard variables.  Check spelling
*>and verify the existence of this variable.
*Examine frequencies of the variables height and weight.
FREQUENCIES VARIABLES = height weight.

*Bivariate Correlation .
CORRELATIONS  all.
```

SPSS automatically comments each session in the background. All syntax, error messages and warning messages are automatically recorded into a journal file (spss.jnl). Each new SPSS session erases the previous journal file. Therefore, rename the journal file to retain it.

Example 1.4 *SPSS Help and Documentation*

Open the SPSS Help Index by selecting Help-**Topics** on the menu bar. First-time SPSS users should review the **Syntax, Command Syntax** and **SPSS basic steps** topics. To obtain contextual help while editing syntax place the cursor in an SPSS keyword and click the **Syntax help** menu bar button ( ). For example, place the cursor in the DESCRIPTIVES keyword and click the **Syntax help** menu bar button to open the complete DESCRIPTIVES command syntax diagram, pictured below.

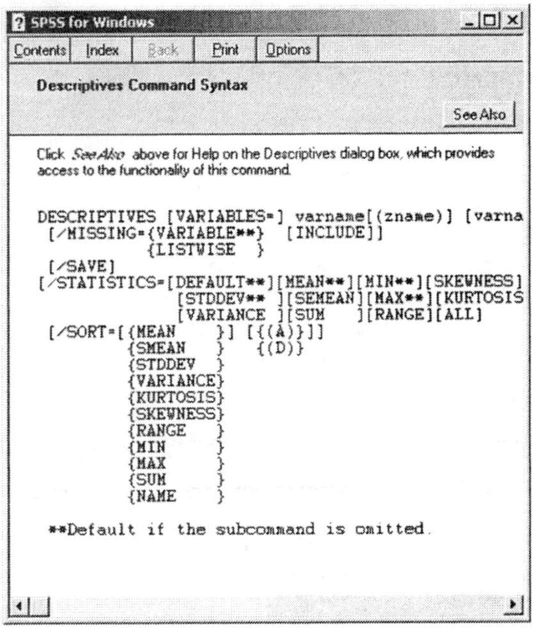

SPSS for Windows

Contents | Index | Back | Print | Options

Descriptives Command Syntax

See Also

Click *See Also* above for Help on the Descriptives dialog box, which provides access to the functionality of this command.

```
DESCRIPTIVES [VARIABLES=] varname[(zname)] [varna
 [/MISSING={VARIABLE**}  [INCLUDE]]
           {LISTWISE  }
 [/SAVE]
[/STATISTICS=[DEFAULT**][MEAN**][MIN**][SKEWNESS]
             [STDDEV**  ][SEMEAN][MAX**][KURTOSIS
             [VARIANCE  ][SUM    ][RANGE][ALL]
 [/SORT=[{MEAN     }] [{(A)}]]
         {SMEAN    }   {(D)}
         {STDDEV   }
         {VARIANCE}
         {KURTOSIS}
         {SKEWNESS}
         {RANGE    }
         {MIN      }
         {MAX      }
         {SUM      }
         {NAME     }

**Default if the subcommand is omitted.
```

Output Help

Right-click a table in the SPSS Output Viewer. Select **Results Coach** in the popup menu.

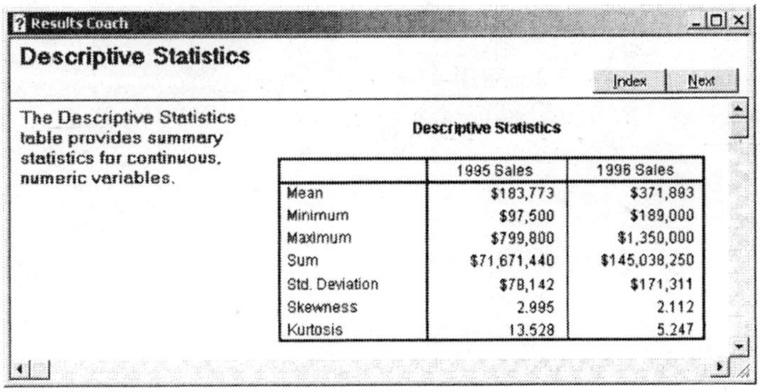

Results Coach

Descriptive Statistics

Index | Next

The Descriptive Statistics table provides summary statistics for continuous, numeric variables.

Descriptive Statistics

	1995 Sales	1996 Sales
Mean	$183,773	$371,893
Minimum	$97,500	$189,000
Maximum	$799,800	$1,350,000
Sum	$71,671,440	$145,038,250
Std. Deviation	$78,142	$171,311
Skewness	2.995	2.112
Kurtosis	13.528	5.247

Syntax Reference

The **SPSS Syntax Reference Guide** is the authoritative source on SPSS Base command rules encountered in this Guide.[2] Its Universals section contains a definitive discussion of SPSS command rules. The Syntax Reference Guide in .pdf format is a handy on-screen reference during programming.

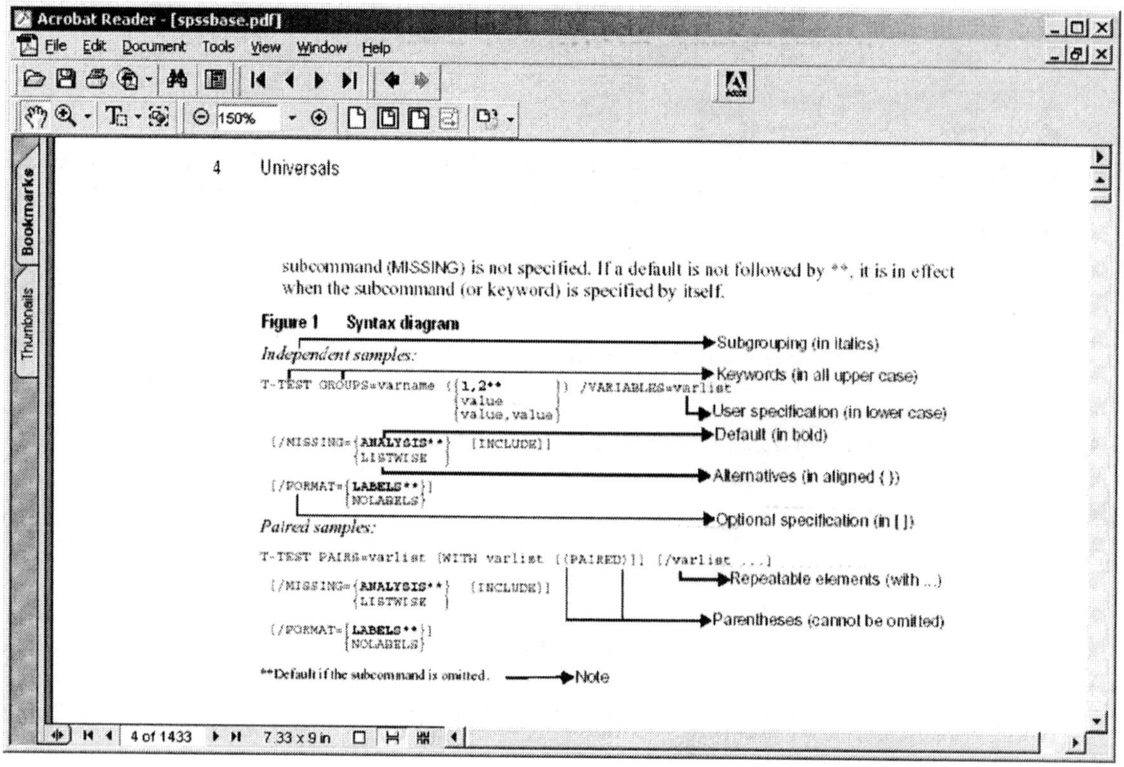

Acrobat Reader - [spssbase.pdf]

File Edit Document Tools View Window Help

4 Universals

subcommand (MISSING) is not specified. If a default is not followed by **, it is in effect when the subcommand (or keyword) is specified by itself.

Figure 1 Syntax diagram

Independent samples:

```
T-TEST GROUPS=varname ({1,2**          }) /VARIABLES=varlist
                        {value          }
                        {value,value    }
  [/MISSING={ANALYSIS**}  [INCLUDE]]
            {LISTWISE  }
  [/FORMAT={LABELS**}]
           {NOLABELS}
```

Paired samples:

```
T-TEST PAIRS=varlist [WITH varlist [(PAIRED)]] [/varlist ...]
  [/MISSING={ANALYSIS**}  [INCLUDE]]
            {LISTWISE  }
  [/FORMAT={LABELS**}]
           {NOLABELS}
```

**Default if the subcommand is omitted.

Subgrouping (in italics)
Keywords (in all upper case)
User specification (in lower case)
Default (in bold)
Alternatives (in aligned { })
Optional specification (in [ ])
Repeatable elements (with ...)
Parentheses (cannot be omitted)
Note

4 of 1433 7.33 x 9 in

[2] SPSS 10.0 Syntax Reference Guide, copyright 1999 by SPSS Inc..

2 SPSS Matrix Manipulation

New syntax editor

This chapter introduces the SPSS matrix processor to support discussion on matrix algebra in the Analyzing Multivariate Data textbook. The contents of this chapter are therefore background material. Readers interested solely in substantive multivariate techniques should skip to Chapter 3 on Regression Analysis.

Example 2.1 *Creating a Matrix statement*

The SPSS matrix processor permits arithmetic manipulation of matrices and composing matrix algebraic programs. An SPSS matrix program consists of matrix 'statements' initiated by the MATRIX command and terminated by the END MATRIX command. The basic format for creating a matrix program is:

```
MATRIX.
[matrix statements].
END MATRIX.
```

The matrix statement COMPUTE M1={1,2;1,2} creates a 2 x 2 matrix named 'M1.' Values of the matrix are enclosed in brackets and separated by commas. A semicolon separates matrix rows.

Ch2sps.sps - SPSS Syntax Editor
File Edit View Analyze Graphs Utilities Run Window Help

```
MATRIX.
COMPUTE M1 ={1,2;1,2}.
PRINT M1.
END MATRIX.
*the PRINT statement creates output showing matrix M1.
```

SPSS Processor is

Keyword
Command terminator
Variable
Matrix specification

A matrix statement can have 6 elements: Keywords, variable names, matrix specifications enclosed in braces({}), operators, functions and the SPSS command terminator (a period). Note that commas separate matrix specification arguments. A semicolon indicates a new row.

Unlike Base SPSS, matrix processor output is not automatically displayed. Use the PRINT statement to display matrix output in the Output Window. Matrix statements can be commented as other SPSS syntax, by a * statement with a period terminator. Alternatively, text between /* and */ anywhere within a statement is ignored by the matrix processor.

```
Run MATRIX procedure:

M1
  1  2
  1  2

------ END MATRIX -----
```

The SPSS matrix processor has some important differences from SPSS Base software. SPSS matrix statements cannot be generated from SPSS Base menus and must be typed directly into syntax files. Another difference is that the Matrix processor will run without a working data file, and ignores the SPSS working data file except where statements like GET or SAVE explicitly read or write SPSS format data files.

Example 2.2 *Defining matrices -- The COMPUTE statement*

Use the COMPUTE statement to create matrices and do much of the work of the matrix program. The basic COMPUTE term is:

COMPUTE target variable = Expression (matrix function; operators and/or functions)

COMPUTE, like most SPSS matrix statements, has an identical name and analogous function to a comparable SPSS command.

Matrix elements are enclosed in braces ({}), with commas separating elements within a matrix row and semicolons separating matrix rows. For example {4,5,6} represents a row vector: [4,5,6]

{4;5;6 } represents a column vector:

$$\begin{bmatrix} 4 \\ 5 \\ 6 \end{bmatrix}$$

{1,2;5,6} represents the 2 x 2 matrix:

$$\begin{pmatrix} 1 & 2 \\ 5 & 6 \end{pmatrix}$$

A 2 x 2, a 3 x 3 and a 2 x 4 matrix can be specified as follows.

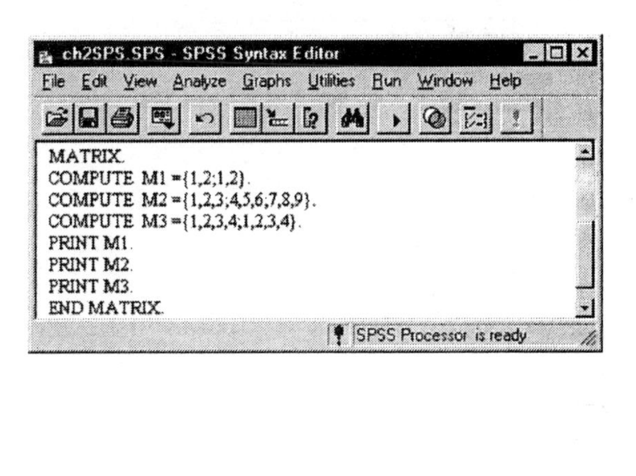


```
Output9 - SPSS Viewer
Output / Matrix / Title / Notes / Text Output

Matrix

Run MATRIX procedure:

M1
  1  2
  1  2

M2
  1  2  3
  4  5  6
  7  8  9

M3
  1  2  3  4
  1  2  3  4

------ END MATRIX -----

SPSS Processor is ready
```

Within the COMPUTE expression you may employ any of the typical arithmetic operators (such as + or *), relational and logical operators (such as LT, GE), Statistical functions (such as SUM or MEAN), cumulative distribution functions, matrix constants, matrix variables, and matrix functions.

COMPUTE N = 10.	Creates a scalar N equal to 10
COMPUTE P = 20 * othervar.	Creates a scalar P equal to 20 * othervar

SPSS matrix math typically requires that matrices be 'conformable', that is, have appropriate shape for the particular mathematical operation to succeed. Addition, subtraction, relational and logical operations require same-sized matrices. Multiplication requires number of columns in the first matrix equal number of rows in the second matrix. Where matrices are not conformable SPSS returns an error message and will not complete the computation. Conformability requirements for Matrix functions are described in the SPSS Base Syntax Guide section on Matrix functions and in the discussion of the Matrix COMPUTE statement.

COMPUTE Matrix3 = Matrix1 * Matrix2
This expression creates Matrix3 as a multiple of Matrix1 and Matrix2. The number of columns in the first-named matrix (Matrix1) must be equal to the number of rows in the second-named matrix (Matrix2). For example, you may multiply a 4x3 matrix by a 3x2 matrix, but may not multiply a 3x2 by a 4x3 matrix.

COMPUTE Matrix3 = Matrix1 + Matrix2   Adds Matrix1 and Matrix2. Matrices must have identical row and column sizes for addition and subtraction.
Relational and logical operators such as GT, LE, EQ, NOT, OR, AND require matrices of same size.

```
MATRIX.
COMPUTE  M5 ={1,2;1,2}.
COMPUTE  M6 ={4,5;6,7}.
COMPUTE M7 = M5 + M6.
PRINT M7.
END MATRIX.
```

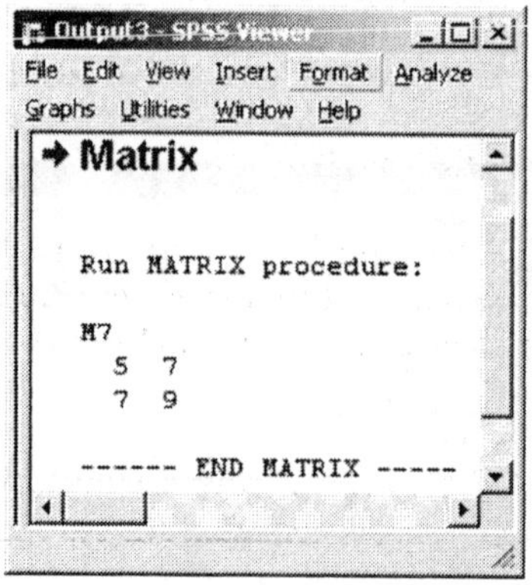

```
Matrix

Run MATRIX procedure:

M7
  5  7
  7  9

------ END MATRIX -----
```

Example 2.3 *Matrix functions, procedures and logical statements*

SPSS matrix algebra functions each take a numeric matrix as an argument. For example, within the COMPUTE statement:

COMPUTE matrix = function (argument1, argument2, argument3)

Taking the function DET (Determinant), COMPUTE DetM = DET(M) takes a single argument, which must be a square matrix (m x m) and returns a scalar which is the determinant of M.

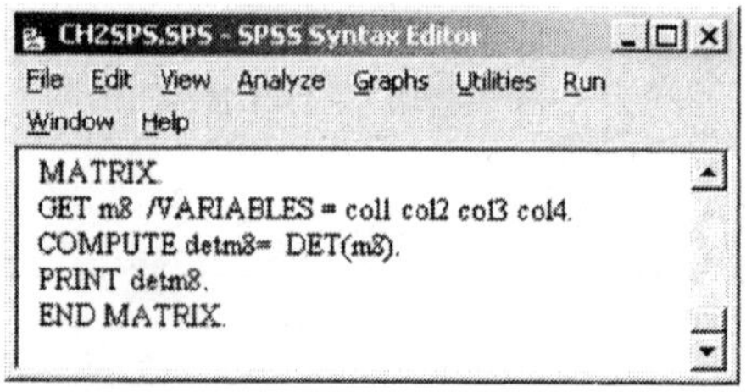

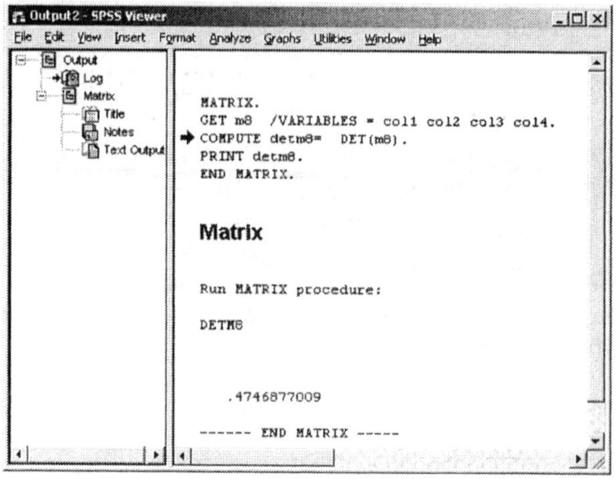

```
MATRIX.
GET m8 /VARIABLES = col1 col2 col3 col4.
COMPUTE detm8= DET(m8).
PRINT detm8.
END MATRIX.
```

```
MATRIX.
GET m8  /VARIABLES = col1 col2 col3 col4.
COMPUTE detm8=  DET(m8).
PRINT detm8.
END MATRIX.

Matrix

Run MATRIX procedure:

DETM8

   .4746877009

------ END MATRIX -----
```

CSUM Column sums COMPUTE Sum1 = CSUM (M1)
CSUM accepts a single argument, in this example the matrix M1. This example returns a row vector (Sum1) with sums of columns of M1

EVAL Eigenvalues of symmetric matrix COMPUTE Eigen1 = EVAL (S_matrix1)
This creates a column vector Eigen1, containing eigenvalues of S_matrix1. EVAL requires that the argument (in this example the matrix S_matrix1) must be symmetrical.

IDENT Create identity matrix COMPUTE Id1 = IDENT (M1)
COMPUTE Id2 = IDENT (M1,M2)
IDENT returns an identity matrix with elements on the main diagonal 1 and off-diagonal elements 0. The first example returns square matrix Id1; the second example returns Id2, with as many rows as M1 and as many columns ad M2.

INV Inverse COMPUTE Minv1 = INV (M1)
INV(M1) creates matrix Minv1, the inverse of M1. INV accepts a single square, nonsingular argument (determinant NE 0). Minv1 has the property that M1* Minv1 = Minv1 * M1 = 1.

MAKE Create a matrix with equal elements COMPUTE matrix1 = MAKE (2,3,5)
MAKE takes 3 scalars as arguments. It creates an Arg1 x Arg2 matrix with all elements equal to Arg3. This example statement creates a 2 X 3 matrix with all elements 5.

MDIAG Diagonal of matrix COMPUTE M2 = MDIAG (Vector1)
Creates square matrix M2 with as many rows and columns as the length or width of Vector1. The diagonal of M2 contains the elements of Vector1. Off-diagonal values are 0.

SVAL Singular values COMPUTE SV1 = SVAL (M1)
This example takes the singular values of M1. SV1 is a column vector with rows equal to the minimum of the rows and columns of M1, containing the singular values of M1 in decreasing order of magnitude.

TRANSPOS Transposition of matrix COMPUTE Trans1 = T (Matrix1)
T is a synonym for TRANSPOS. Computes Trans1 as the transpose of Matrix1 with rows and columns interchanged. If Matrix1 is an n x m matrix, then Trans1 is an m x n matrix. Note also that the SPSS Data Editor menu **Data** transposes data.

Matrix procedures and Logical statements

The CALL statement invokes matrix procedures. Matrix procedures are similar to Matrix functions except that procedures can return more than a single value result.

> SETDIAG (M,V) sets the main diagonal of a matrix M to the values of a vector, V, modifying the existing matrix without creating a copy of it.
>
> SVD (M,var1,var2,var3) performs a singular Value decomposition of a matrix M, with results assigned to matrices corresponding to var1, var2 and var3.

SPSS Matrix language includes a variety of logical control structure statements (also contained in Base SPSS).

> DO IF followed by a logical expression determines criteria for execution of matrix statements on particular matrices.
>
> LOOP defines the beginning of a block of statements to be repeatedly run until a DO IF criterion is satisfied or a BREAK statement is encountered.
>
> Other control structure clauses and keywords include BY, ELSE IF, ELSE, LOOP IF, END IF and END LOOP.

3 Regression Analysis Using SPSS

3.1 **Bivariate regression predicting weight from height**
3.2 **Modeling market valuation**
3.3 **Diagnosing and dealing with heteroskedasticity**

The SPSS REGRESSION command performs a wide range of statistical and diagnostic tests of a regression model. REGRESSION includes a variety of stepwise regression and model selection methods.

Example 3.1 *Heights and weights of 20 women*

This example demonstrates using a scatterplot to visually assess data relationships and a bivariate regression to predict weight from height. The example follows discussion in textbook Chapter 3, Section 3.

Example 3.1.A *Rename variables col1, col2 to Weight and Height*

Variables in women.sav as copied from the CD-Rom accompanying Analyzing Multivariate Data are col1 and col2. The actual names of the women.sav variables are Height and Weight:

Variable Name	Description
Height	Height in inches
Weight	body weight in pounds

Rename variables col1 and col2 to Height and Weight by directly text editing the Name column within the **Variable View** of the Data Editor. Renaming variables is a convenience to make output easier to understand. It does not affect statistical results.

Women.sav - SPSS Data Editor

File Edit View Data Transform Analyze Graphs Utilities Window Help

	Name	Type	Width	Decimals
1	height	Numeric	2	0
2	weight	Numeric	3	0
3				

Data View | Variable View

SPSS Processor is ready

Example 3.1.B *Create a bivariate scatter plot to visually assess the relationship of height to weight*

The **Graphs-Scatter** menu path generates the **Scatterplot** dialog box. Click the **Simple** and **Define** buttons to display the **Simple: Scatterplot** dialog box. Assign variables to X and Y axes. Click the **OK** button to run or click the **Paste** button to edit the command in a syntax file.

Scatterplot: Simple | Matrix | Overlay | 3-D | Define | Cancel | Help

Simple Scatterplot: Y Axis: height | X Axis: weight | Set Markers by: | Label Cases by: | Template | Use chart specifications from: | File... | Titles... | Options... | OK | Paste | Reset | Cancel | Help

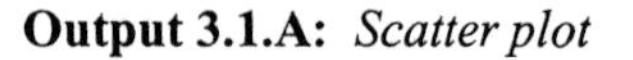

Output 3.1.A: *Scatter plot*

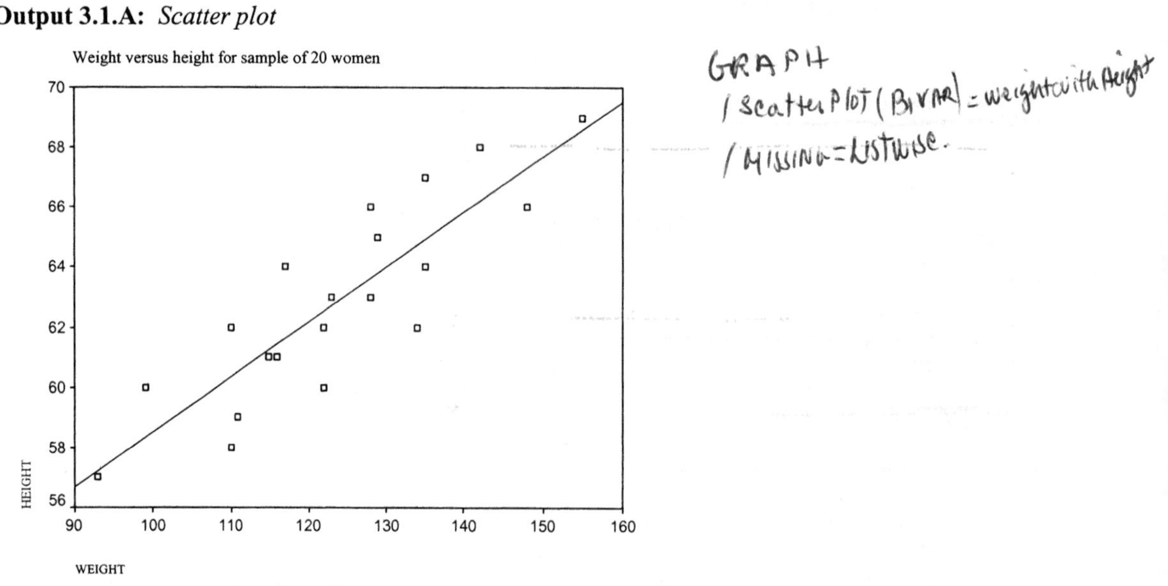

The scatter plot reveals a strong correlation between weight and height, also suggested by the inclination of the superimposed regression line. Close clustering of data points to the line indicates there will be good fit to data in the regression model.

Example 3.1.C *Build a regression model predicting weight from height*

The SPSS REGRESSION command can be run interactively from the **Analyze-Regression-Linear** menu path, which produces the **Linear Regression** dialog box. Assign Height as the independent variable and Weight as dependent variable. Syntax can be pasted from the dialog boxes into an SPSS Syntax file for further editing.

Syntax can be pasted from the dialog boxes into an SPSS Syntax file for further editing.

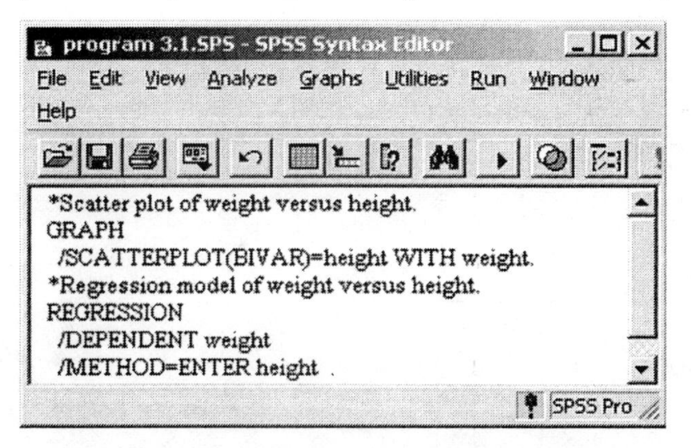

Two subcommands are required for SPSS REGRESSION:

- /DEPENDENT identifies the dependent variable.
- /METHOD specifies the method for selecting independent variables, and lists independent variables.

Output 3.1.B *Regression model*

The Model Summary table lists key regression statistics. The R statistic of .867 indicates height and weight have relatively high correlation. The regression equation explains 75% of the variability of the data (R Square =.752). The adjusted R square is an indicator of fit in models containing multiple independent variables.

Model Summary

Model	R	R Square	Adjusted R Square	Std. Error of the Estimate
1	.867[a]	.752	.738	7.97

a. Predictors: (Constant), HEIGHT

The ANOVA table contains the statistical test of significance of the model. The substantial F statistic of 54.5, and significance of .000 lead us to be confident there is a very strong relationship between height and weight.

ANOVA[b]

Model		Sum of Squares	df	Mean Square	F	Sig.
1	Regression	3463.460	1	3463.460	54.526	.000[a]
	Residual	1143.340	18	63.519		
	Total	4606.800	19			

a. Predictors: (Constant), HEIGHT

b. Dependent Variable: WEIGHT

The Coefficients table estimates individual parameters (Betas) and tests of whether parameters are significantly different from zero. The Beta for height (4.095) predicts that every inch of height gain is associated with 4.1 pounds greater body weight. The ANOVA table above has already confirmed that the relationship of height to weight is significantly different from zero, so the t statistic and Significance are redundant.

Coefficients[a]

		Unstandardized Coefficients		Standardized Coefficients		
Model		B	Std. Error	Beta	t	Sig.
1	(Constant)	-133.764	34.899		-3.833	.001
	HEIGHT	4.095	.555	.867	7.384	.000

a. Dependent Variable: WEIGHT

Example 3.2 *Regression Analysis of Leslie Salt data*

Example 3.2 demonstrates a regression model, how to compute diagnostic statistics and how to compute additional variables in order to compensate for data skewness or for interaction effects in the regression model. The data set, Leslie_salt.sav, contains six variables, col1 through col6. These are renamed to the following:

Variable Name	Description
Price	Sale price in $000 per acre
County	San Mateo = 0, Santa Clara = 1
Eleva	Average elevation in feet above sea level
Date	Date of sale counting backward from current time (in months)
Flood	Subject to flooding by tidal action =1; otherwise =0
Distance	Distance in miles from Leslie property

All analyses can be run from menus or pasted into a syntax file.

Program 3.2.SPS - SPSS Syntax Editor

File Edit View Analyze Graphs Utilities Run Window Help

```
*SPSS PROGRAM 3.2: Two models of property value using Leslie Salt data (leslie_salt.sav).

*Histogram of PRICE.
GRAPH
 /HISTOGRAM(NORMAL)=price .
*Compute Natural log of Price to reduce skewness of the dependent variable.
COMPUTE logprice = Ln(price) .
EXECUTE .

*Regression model I (equation 3.17 in textbook).
*The subcommands specify regression statistics, Durbin-Watson test statistic,
*Collinearity diagnostics and a table of diagnostics for influential observations.
REGRESSION
 /STATISTICS COEFF OUTS R ANOVA COLLIN TOL
 /DEPENDENT logprice
 /METHOD=ENTER date distance eleva flood
 /RESIDUALS DURBIN
 /CASEWISE= PLOT(resid) all sresid LEVER COVRATIO sDFIT.

*Regression diagnostics, including DFBETAs table.
REGRESSION
 /DEPENDENT logprice
 /METHOD=ENTER date distance eleva flood
 /CASEWISE= PLOT(sdresid) all dfbeta
 /SAVE SRESID .

*Plot Studentized residuals versus elevation.
GRAPH
 /SCATTERPLOT(BIVAR)=eleva WITH sre_1.

*Compute a variable multiplying county code times elevation (see discussion p. 66).
COMPUTE c_x_e = county*elevatio .
EXECUTE .

*Regression model II (equation 3.37 in the textbook).
REGRESSION
 /STATISTICS COEFF OUTS R ANOVA COLLIN TOL
 /DEPENDENT LOGPRICE
 /METHOD=ENTER DATE DISTANCE ELEVA FLOOD COUNTY c_x_e.
```

SPSS Processor is ready

Example 3.2.A *Define a histogram and create a new independent variable*

First define a histogram of the dependent variable Price. Open the histogram dialog box from the menu path **Graph-Histogram:**

The histogram demonstrates skewed distribution of the dependent variable, Price.

SPSS Output 3.2.A

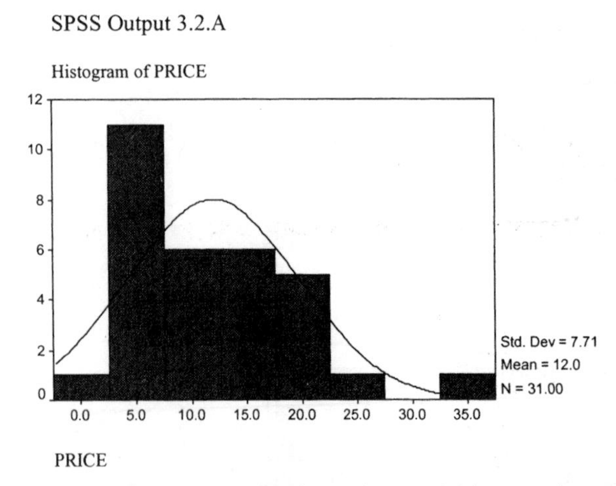

To reduce skewness in the dependent variable use a logarithmic transformation. Select the menu path **Transform-Compute** in the Data Editor to generate the **Compute Variable** dialog box. Transfer LN(numexpr), the numeric expression to compute LOGPRICE, from the Functions list and secondly transfer price from the Variables list. Note that the **Transform** menu is only available from the Data Editor. The **Transform** menu does not appear as a menu item in SPSS Syntax or Viewer files.

Example 3.2.B *Regression Model I*

In this example define a regression model of home pricing, here labeled Model I. Begin building the Leslie Salt Regression model in a manner similar to Example 3.1, with the menu path **Analyze-Regression-Linear**.

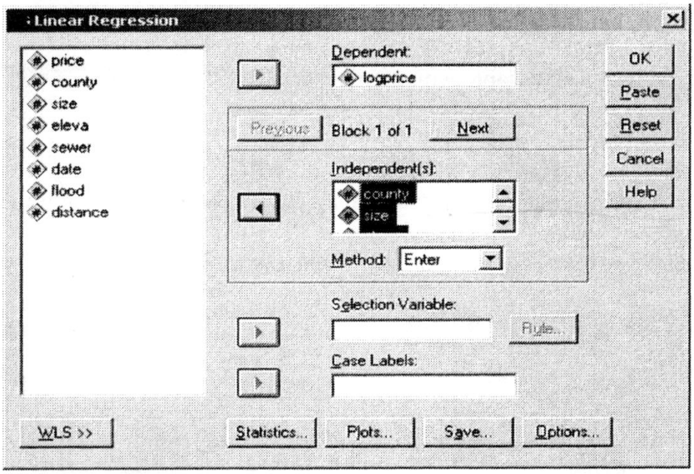

Click the **Statistics** button to display the **Linear Regression: Statistics** dialog box for defining regression diagnostics:

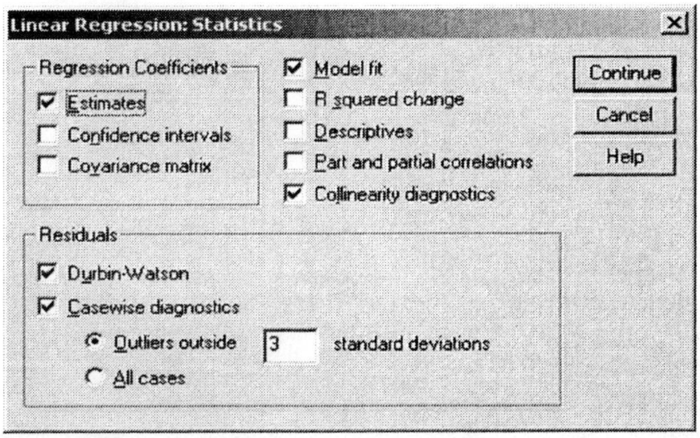

The **Estimates** and **Model fit** checkboxes specify default Regression output, including a Model Summary table, an ANOVA table and a Coefficients estimates table. Click the **Collinearity diagnostics** check box to assess multicollinearity among independent variables. Click on the **Casewise Diagnostics** checkbox to obtain a diagnostics table of standardized residuals. Click on the **Durbin-Watson** box to obtain the Durbin-Watson statistic.

SPSS REGRESSION automatically creates 19 temporary residuals variables, most of which can be plotted or displayed in tables for residuals analysis. Several of these temporary residuals variables can be added to the Casewise Diagnostics table by editing the /CASEWISE subcommand. The following syntax creates a Casewise Diagnostics table with columns for unstandardized residuals, studentized residuals, centered leverage values, covariance ratios and standardized DFFITs:

```
/CASEWISE= PLOT(resid) all sresid lever covratio  sdfit
```

The following syntax creates a Casewise Diagnostics table with dfBETAs:

```
/CASEWISE= PLOT(sdresid)  all dfbeta
```

Click on the **Save** button of the **Linear Regression** dialog box to display the **Linear Regression: Save** dialog box. Then click the **Studentized residuals** check box. For the Leslie Salt regression analysis we will save Studentized residuals into the working data file and plot them against Elevation.

Linear Regression: Save

Predicted Values: Unstandardized; Standardized; Adjusted; S.E. of mean predictions

Residuals: Unstandardized; Standardized; Studentized (checked); Deleted; Studentized deleted

Continue | Cancel | Help

Distances: Mahalanobis; Cook's; Leverage values

Influence Statistics: DfBeta(s); Standardized DfBeta(s); DfFit; Standardized DfFit; Covariance ratio

Prediction Intervals: Mean; Individual; Confidence Interval: 95 %

Save to New File: Coefficient statistics | File...

Export model information to XML file | Browse

Output 3.2.B *Results for Regression Model I*

The Model Summary Table demonstrates that the regression equation explains 78% of the variability of the data (R Square =.781). The adjusted R square of .747 is nearly equivalent to R Square, indicating the model has good fit. The Durbin-Watson statistic is included in the Model Summary. For interpretation of this estimate of autocorrelation among residuals, see p. 61 of Text.

Model Summary[b]

Model	R	R Square	Adjusted R Square	Std. Error of the Estimate	Durbin-Watson
1	.883[a]	.781	.747	.361	2.376

a. Predictors: (Constant), FLOOD, DATE, elevation, DISTANCE

b. Dependent Variable: LOGPRICE

The ANOVA table contains a substantial F statistic of 23.121 and significance of .000.

ANOVA[b]

Model		Sum of Squares	df	Mean Square	F	Sig.
1	Regression	12.032	4	3.008	23.121	.000[a]
	Residual	3.382	26	.130		
	Total	15.414	30			

a. Predictors: (Constant), FLOOD, DATE, elevation, DISTANCE

b. Dependent Variable: LOGPRICE

The Coefficients table suggests that all four parameters of Model I are significantly different from zero, and therefore contribute to the explanatory power of the model. The model predicts that Date and Distance contribute positively to logprice, while Flood has a negative relationship.

Coefficients[a]

Model		Unstandardized Coefficients		Standardized Coefficients	t	Sig.	Collinearity Statistics	
		B	Std. Error	Beta			Tolerance	VIF
1	(Constant)	2.824	.214		13.223	.000		
	DATE	1.857E-02	.003	.635	6.902	.000	.996	1.004
	DISTANCE	5.908E-02	.017	.374	3.575	.001	.772	1.296
	elevation	7.459E-02	.017	.453	4.435	.000	.808	1.237
	FLOOD	-.779	.201	-.406	-3.868	.001	.765	1.307

a. Dependent Variable: LOGPRICE

Collinearity Diagnostics[a]

Model	Dimension	Eigenvalue	Condition Index	Variance Proportions				
				(Constant)	DATE	DISTANCE	elevation	FLOOD
1	1	3.333	1.000	.01	.01	.02	.02	.02
	2	1.043	1.788	.00	.00	.03	.11	.35
	3	.351	3.083	.00	.00	.39	.25	.63
	4	.212	3.965	.02	.28	.40	.49	.00
	5	6.158E-02	7.357	.97	.70	.16	.14	.00

a. Dependent Variable: LOGPRICE

The Collinearity diagnostics table contains three types of indicators: eigenvalues of the cross-products matrix, a Condition Index and tolerances for individual variables. Where tolerances are less than .01 multicollinerity problems are indicated. The results of the table below suggest there is not a problem with multicollinearity in the Leslie Salt regression Model I. This is another indication individual coefficients are significantly different from zero.

Output 3.2.C *Diagnostics for Influential observations*

The Residuals Statistics table summarizes the range and mean of residuals. These can indicate outliers and significant variability in residuals.

Residuals Statistics[a]

	Minimum	Maximum	Mean	Std. Deviation	N
Predicted Value	.834	3.225	2.259	.633	31
Std. Predicted Value	-2.251	1.525	.000	1.000	31
Standard Error of Predicted Value	7.262E-02	.249	.137	4.878E-02	31
Adjusted Predicted Value	.961	3.447	2.275	.652	31
Residual	-.510	1.003	-6.45E-17	.336	31
Std. Residual	-1.414	2.781	.000	.931	31
Stud. Residual	-1.546	3.036	-.019	1.026	31
Deleted Residual	-.610	1.195	-1.54E-02	.412	31
Stud. Deleted Residual	-1.591	3.705	.009	1.110	31
Mahal. Distance	.248	13.354	3.871	3.435	31
Cook's Distance	.000	.353	.049	.077	31
Centered Leverage Value	.008	.445	.129	.114	31

a. Dependent Variable: LOGPRICE

The first column of the Casewise Diagnostics table contains case numbers, the second the standardized residual. Column 3 contains the independent variable value. Column 4 lists Studentized residuals, which are the residual divided by an estimate of its standard deviation. Where Studentized residuals are large a case is likely more influential. Column 5 lists Leverage Values, which indicate the influence of particular cases on coefficients. Column 6, COVRATIO, contains ratios of the

Casewise Diagnostics[a]

Case Number	Std. Residual	LOGPRICE	Stud. Residual	Centered Leverage Value	COVRATIO	Std. DFFIT	DFBETA (Constant)	FLOOD	DATE	elevation	DISTANCE
1	-.475	1.5	-.530	.166	1.438	-.260	2.451E-02	-5.21E-03	5.008E-04	-1.57E-03	1.188E-03
2	2.781	2.4	3.036	.128	.163	1.621	-7.715E-02	-4.19E-02	-2.94E-03	-5.46E-03	-4.825E-03
3	-.842	.5	-1.002	.262	1.417	-.648	5.492E-02	-7.89E-02	9.734E-04	1.022E-03	-7.290E-04
4	-1.080	1.6	-1.301	.278	1.259	-.884	7.679E-02	8.632E-02	1.169E-03	2.819E-03	-1.042E-02
5	1.027	1.6	1.239	.280	1.304	.843	-8.991E-02	8.557E-02	-1.05E-03	9.731E-04	5.341E-03
6	-.506	1.2	-.552	.126	1.363	-.236	-1.023E-02	5.689E-03	3.305E-04	2.155E-03	2.276E-03
7	-.331	1.7	-.348	.059	1.308	-.109	-1.045E-02	1.511E-03	6.443E-05	6.968E-04	1.268E-03
8	-.304	1.8	-.319	.056	1.309	-.098	-1.112E-02	1.351E-03	3.216E-05	6.298E-04	1.165E-03
9	-.698	3.0	-.966	.445	1.939	-.922	6.062E-02	-4.20E-02	-1.56E-05	-1.46E-02	-5.256E-04
10	-1.414	1.2	-1.546	.132	.899	-.705	-9.178E-02	2.332E-02	1.341E-04	8.477E-03	7.176E-03
11	-.399	1.5	-.437	.131	1.402	-.190	-2.648E-02	6.573E-03	2.829E-05	2.391E-03	2.028E-03
12	-.007	1.9	-.007	.068	1.352	-.002	-3.213E-04	4.824E-05	1.647E-07	1.996E-05	2.782E-05
13	-.109	2.1	-.113	.038	1.306	-.031	-3.790E-03	3.551E-04	-9.33E-07	1.152E-04	3.459E-04
14	-.373	2.5	-.383	.022	1.250	-.090	-3.411E-04	3.026E-03	-4.37E-06	-8.27E-04	-2.237E-04
15	.658	3.0	.688	.053	1.213	.208	-5.429E-03	-1.56E-03	1.680E-05	2.610E-03	5.670E-04
16	-.572	2.5	-.593	.036	1.220	-.158	2.601E-03	3.403E-03	-1.01E-05	-1.78E-03	-5.094E-04
17	-.043	2.6	-.044	.023	1.287	-.010	2.147E-05	3.891E-04	-4.53E-07	-9.74E-05	-3.594E-05
18	.690	2.7	.704	.008	1.152	.143	5.711E-03	-1.01E-02	-1.62E-06	4.478E-04	4.127E-04
19	-.841	2.5	-.891	.078	1.170	-.312	1.322E-02	9.037E-04	-2.62E-05	-4.24E-03	-1.207E-03
20	.297	2.7	.305	.024	1.267	.074	-5.705E-04	-2.99E-03	2.805E-06	6.946E-04	3.212E-04
21	-.818	2.5	-1.011	.313	1.520	-.734	1.265E-02	8.036E-02	-2.67E-05	2.365E-03	-1.027E-02
22	1.093	2.9	1.116	.009	.993	.233	1.947E-02	-1.98E-02	1.040E-04	-8.76E-05	6.800E-04
23	1.406	2.8	1.452	.031	.850	.385	4.631E-02	-3.84E-02	1.394E-04	-3.55E-03	3.925E-04
24	-.177	1.8	-.200	.184	1.540	-.103	-8.825E-04	-1.40E-02	-3.65E-05	6.209E-05	-3.707E-04
25	-.825	1.4	-.948	.209	1.345	-.533	-3.031E-02	-9.47E-02	-3.01E-04	-1.35E-04	3.094E-03
26	1.990	3.6	2.070	.043	.535	.636	5.516E-02	-4.99E-02	8.290E-04	8.226E-04	4.429E-03
27	.287	2.9	.296	.030	1.276	.075	1.015E-02	-5.48E-03	1.195E-04	2.002E-05	2.352E-04
28	-.587	2.7	-.633	.109	1.311	-.254	-2.107E-02	2.709E-02	-2.99E-04	9.492E-04	-2.566E-03
29	-.261	3.1	-.281	.109	1.396	-.112	-1.823E-02	5.284E-03	-2.59E-04	-6.02E-05	-2.084E-04
30	.816	2.7	1.047	.359	1.613	.842	9.993E-02	.118	1.449E-03	4.768E-04	-4.011E-03
31	-.382	3.1	-.433	.189	1.508	-.227	-4.316E-02	1.200E-02	-5.28E-04	1.023E-03	-4.119E-05

a. Dependent Variable: LOGPRICE

determinant of the covariance matrix with a single case deleted to the determinant of the full covariance matrix of all cases. Values near one indicate the case has little effect on precision of estimates. Column 7 contains standardized DFFITs, which measure how much fitted values change when each particular observation is excluded. DFBETAS columns measure the difference between the regression coefficient estimates with and without each observation (see textbook discussion p. 63 and Table 3.9). We assume that no one observation has unusual influence on parameter estimates because the DFBETA scores are all extremely small.

Output 3.2.D *Results of Regression Model II with an independent variable (c_x_e) accounting for interaction effects*

The steps for creating Regression Model II are identical to the steps to create Regression Model I (Example 3.2). Therefore syntax basis for Output 3.2.D is not explained.

Model Summary

Model	R	R Square	Adjusted R Square	Std. Error of the Estimate
1	.925[a]	.857	.821	.304

a. Predictors: (Constant), C_X_E, DATE, FLOOD, DISTANCE, COUNTY, elevation

ANOVA[b]

Model		Sum of Squares	df	Mean Square	F	Sig.
1	Regression	13.202	6	2.200	23.876	.000[a]
	Residual	2.212	24	9.216E-02		
	Total	15.414	30			

a. Predictors: (Constant), C_X_E, DATE, FLOOD, DISTANCE, COUNTY, elevation

b. Dependent Variable: LOGPRICE

Coefficients[a]

		Unstandardized Coefficients		Standardized Coefficients			Collinearity Statistics	
Model		B	Std. Error	Beta	t	Sig.	Tolerance	VIF
1	(Constant)	1.488	.417		3.566	.002		
	DATE	1.807E-02	.003	.618	6.536	.000	.668	1.496
	DISTANCE	.121	.025	.767	4.919	.000	.246	4.067
	elevation	.320	.073	1.944	4.394	.000	.031	32.733
	FLOOD	-.303	.221	-.158	-1.373	.182	.451	2.219
	COUNTY	1.293	.394	.893	3.284	.003	.081	12.360
	C_X_E	-.266	.076	-1.776	-3.484	.002	.023	43.474

a. Dependent Variable: LOGPRICE

Example 3.3 *Diagnosing and dealing with heteroskedasticity*

This exercise demonstrates how to improve model fit by detecting and resolving problems of heteroskedasticity. The working dataset is lost_days.sav. All steps for the Lost Days analysis can be completed from menus in manner similar to previous exercises of this chapter. Syntax is similar to previous regression analyses with a few exceptions. The regression analysis of dependent variable Lostd (lost days) is identical to previous examples, except for addition of the /SAVE RESID subcommand to save residuals into the working data file. A bivariate scatter plot similar to Example 3.2.A is run, of residual scores versus the independent variable Size. When significant heteroskedasticity is revealed by the scatterplot, a new dependent variable is computed in order to compensate (see the command 'COMPUTE lostpc' in the syntax file displayed below).

heteroskedastic - a set of random variables that have different variances.
homoskedastic - a set of variables all having the same variance

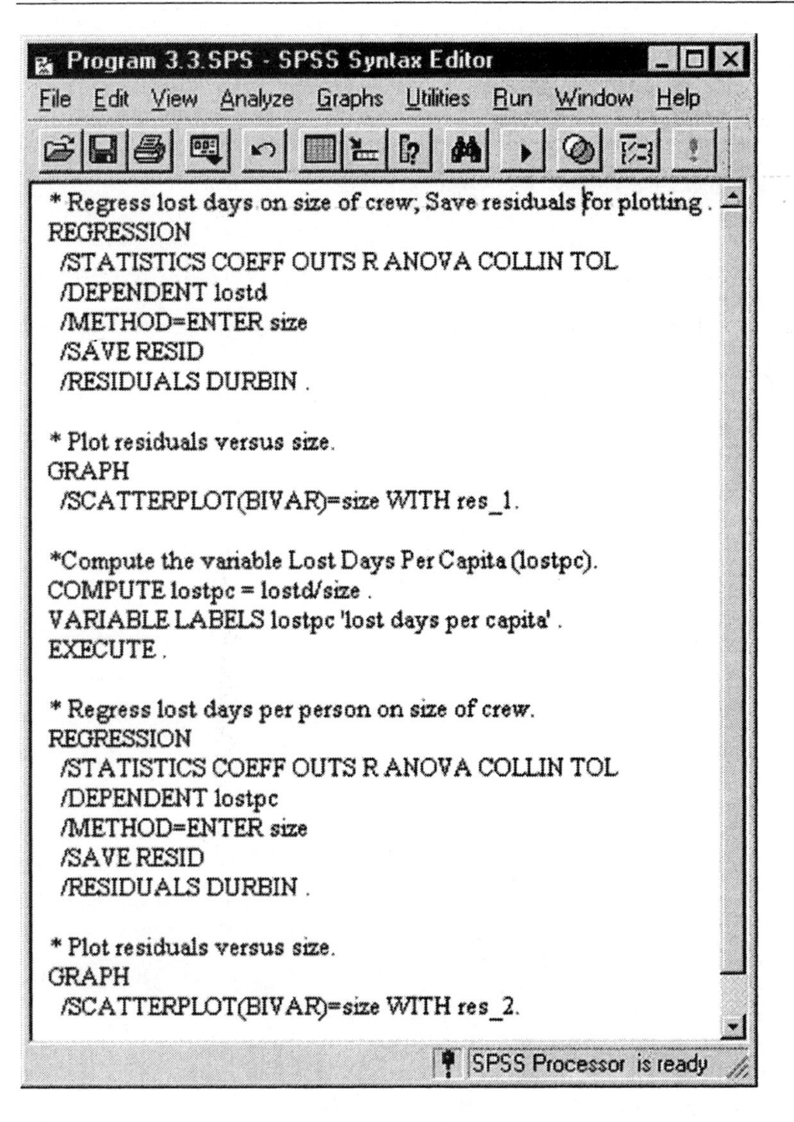


```
* Regress lost days on size of crew; Save residuals for plotting .
REGRESSION
  /STATISTICS COEFF OUTS R ANOVA COLLIN TOL
  /DEPENDENT lostd
  /METHOD=ENTER size
  /SAVE RESID
  /RESIDUALS DURBIN .

* Plot residuals versus size.
GRAPH
  /SCATTERPLOT(BIVAR)=size WITH res_1.

*Compute the variable Lost Days Per Capita (lostpc).
COMPUTE lostpc = lostd/size .
VARIABLE LABELS lostpc 'lost days per capita' .
EXECUTE .

* Regress lost days per person on size of crew.
REGRESSION
  /STATISTICS COEFF OUTS R ANOVA COLLIN TOL
  /DEPENDENT lostpc
  /METHOD=ENTER size
  /SAVE RESID
  /RESIDUALS DURBIN .

* Plot residuals versus size.
GRAPH
  /SCATTERPLOT(BIVAR)=size WITH res_2.
```

Output 3.3.A *Results from regression analysis of lost days on work crew size*

Model Summary[b]

Model	R	R Square	Adjusted R Square	Std. Error of the Estimate	Durbin-Watson
1	.633[a]	.401	.392	33.98	1.841

a. Predictors: (Constant), SIZE

b. Dependent Variable: LOSTD

Residuals Statistics[a]

	Minimum	Maximum	Mean	Std. Deviation	N
Predicted Value	-4.30	89.86	40.42	27.59	67
Residual	-61.74	88.82	1.80E-15	33.72	67
Std. Predicted Value	-1.621	1.792	.000	1.000	67
Std. Residual	-1.817	2.614	.000	.992	67

a. Dependent Variable: LOSTD

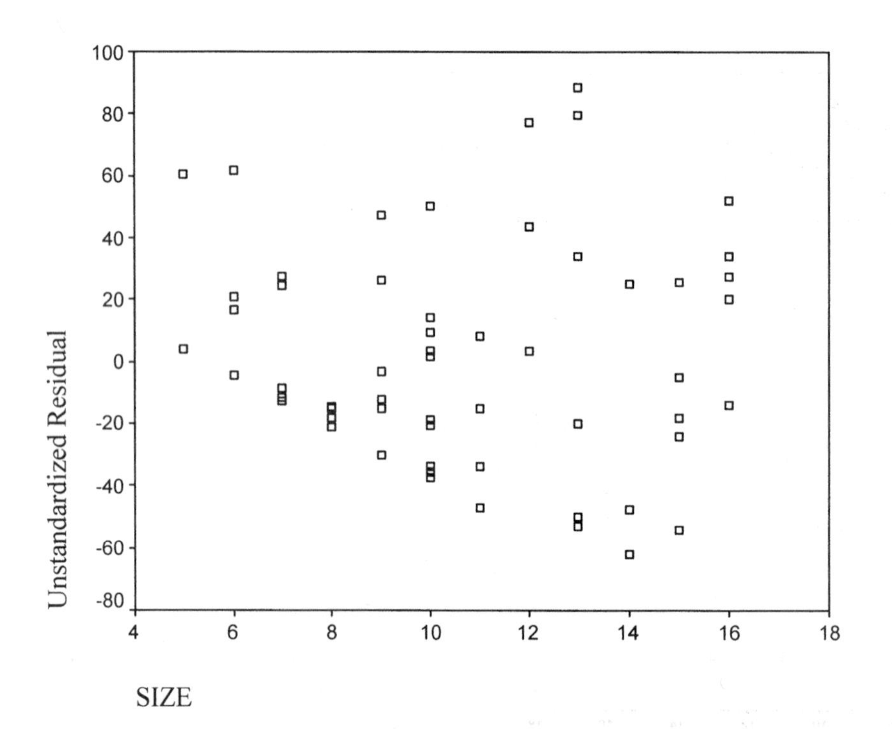

The scatterplot of residuals against crew size indicates heteroskedasticity.

Output 3.3.B *Results from regression of Lost Days Per Capita (Lostpc) on work crew size*

Model Summary[b]

Model	R	R Square	Adjusted R Square	Std. Error of the Estimate	Durbin-Watson
1	.375[a]	.141	.128	3.298	1.969

a. Predictors: (Constant), SIZE

b. Dependent Variable: lost days per capita

ANOVA[b]

Model		Sum of Squares	df	Mean Square	F	Sig.
1	Regression	115.943	1	115.943	10.658	.002[a]
	Residual	707.122	65	10.879		
	Total	823.065	66			

a. Predictors: (Constant), SIZE

b. Dependent Variable: lost days per capita

Coefficients[a]

Model		Unstandardized Coefficients		Standardized Coefficients	t	Sig.	Collinearity Statistics	
		B	Std. Error	Beta			Tolerance	VIF
1	(Constant)	-.663	1.349		-.491	.625		
	SIZE	.411	.126	.375	3.265	.002	1.000	1.000

a. Dependent Variable: lost days per capita

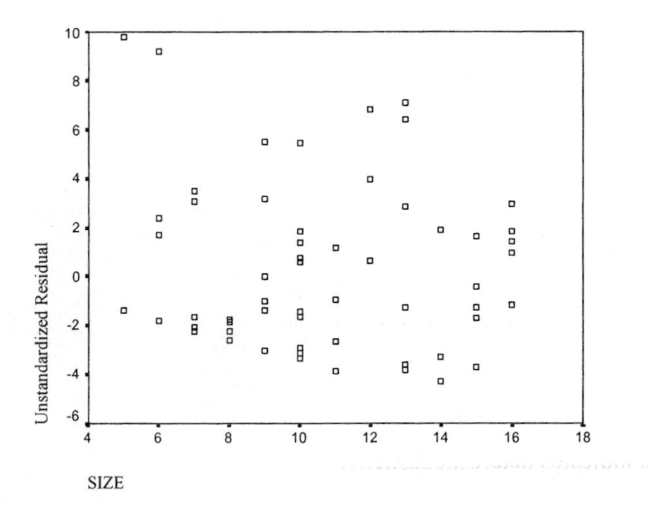

The scatterplot suggests that using 'Lost Days Per capita' as an dependent variable reduces heteroskedasticity.

4 Principal Components Analysis Using SPSS

SPSS FACTOR is a general command used to perform principal components analysis in this chapter, and demonstrate exploratory factor analysis in Chapter 5.

Example 4.1 *Illustrative example*

The data set for this example is pca_example.sav. Change variable names in the Data Editor **Variable View** from col1, col2, and col3 to x1, x2 and x3 to match Section 4.2 of the textbook.

Use the **Graph** menu to create a scatterplot to visually examine the relationship of variables X1 by X2.

Follow the SPSS menu path **Analyze-Data Reduction-Factor** to invoke the **Factor Analysis** Dialog box. Select variables by highlighting the left-hand source list. Click the intervening arrow marker to transfer them to the **Variables** list for analysis.

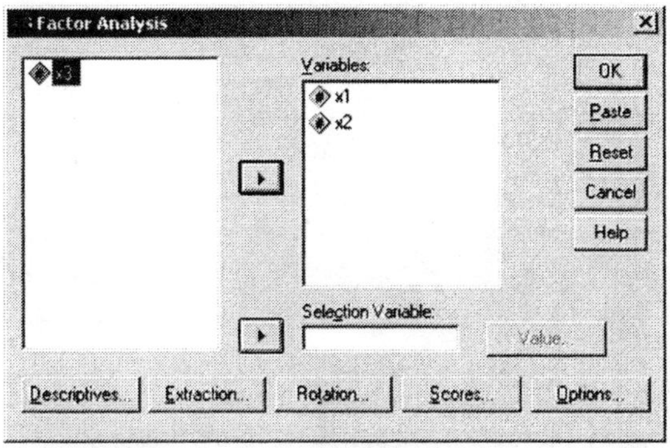

In the example above variables X1 and X2 have already been transferred to the **Variables** list and X3 is highlighted. This generates default FACTOR syntax for principal components analysis:

```
FACTOR
/VARIABLES=varlist.
```

Click the **Descriptives** button to open the **Factor Analysis: Descriptives** dialog box.
Click the Coefficients checkbox to include a Correlation matrix in the output.

Click **Continue** to return to the main **Factor Analysis** dialog box.
Click the **Scores** button to open the **Factor Analysis: Factor Scores** dialog box.

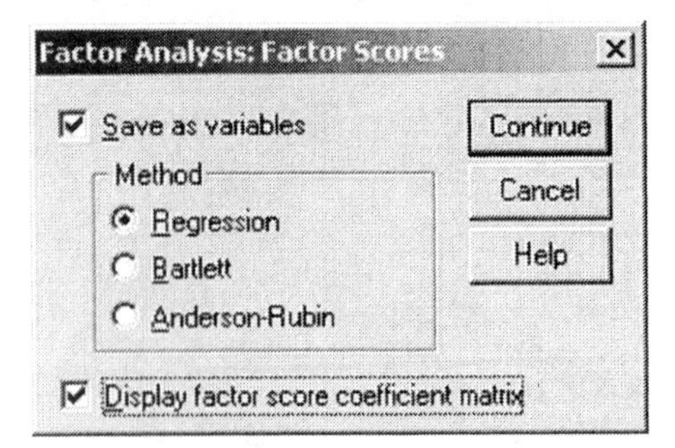

Click the **Save as variables** checkbox to specify that principal component scores will be saved as variables to the working data file. Saved principal component scores will be automatically denoted by the default names fac1_1, fac2_1, and so on. The **Display factor score coefficient matrix** checkbox specifies that a matrix of standardized factor coefficients will be displayed output (the FSCORE syntax option).
Click **Continue** to return to the main **Factor Analysis** dialog box.

You may click **OK** to run the analysis, or click the "Paste" button to transfer the completed commands to the syntax file, shown on the following page with added explanatory comments.

Pasted Factor Analysis syntax

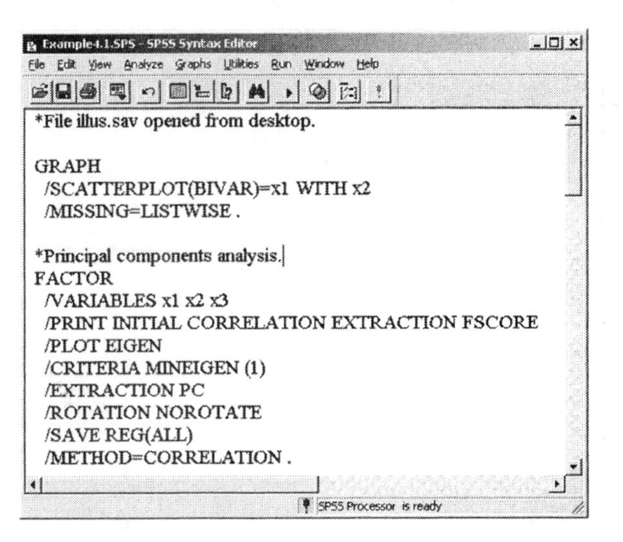

Summary of pasted syntax

- The /VARIABLES subcommand is required, and lists all variables within the factor procedure.
- The /PRINT subcommand determines the format for displaying results.
    - Default options INITIAL and EXTRACTION print tables showing the proportion of variance accounted for by the principal components, eigenvalues, percentage of variance explained by each factor and factor loadings.
    - The CORRELATION option was added from the **Factor Analysis:Descriptives** dialog box, and prints a correlation matrix.
    - The FSCORE option was added from the **Factor Analysis:Factor Scores** dialog box, and will display a matrix of standardized factor scores.
- The /CRITERIA subcommand and option MINEIGEN(1) specifies a minimum eigenvalue level of 1 for factor extraction, default for principal components analysis.
- The /EXTRACTION subcommand defines the solution procedure. The default option, PC, runs the Principal Components procedure.
- The /ROTATION subcommand specifies whether to rotate the factor solution. For principal components the default is NOROTATE. In Chapter 5, we discuss the different subcommands for rotating a factor solution (for example VARIMAX rotations).
- The /SAVE REG(ALL) subcommand adds principal component scores as new variables.
- The /METHOD subcommand specifies whether factor scores are calculated from correlations or covariances. Default is CORRELATION.

Text editing the syntax file

Two FACTOR command options cannot be defined from menus, but must be typed directly into the syntax file for this analysis.

1. The /DIAGONAL subcommand specifies diagonal values as 1 for principal components analysis. Syntax for setting the diagonal values is /DIAGONAL *valuelist,* where *valuelist* is a list of values equivalent to the number of variables in the analysis. Where a value is repeated SPSS accepts the argument n* to indicate number of repetitions. The notation is /DIAGONAL 3*1 for this 3 variable analysis.
2. The FACTORS(n) option of the /CRITERIA subcommand can be directly typed into a syntax file to define number of components to be extracted. FACTORS (3) is the specification for this analysis.

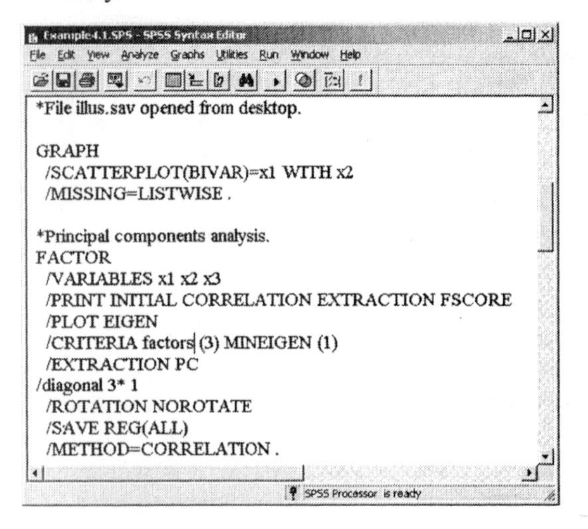

Output 4.1.A *Scatterplot of variables X1 and X2 in the Illustrative data set*

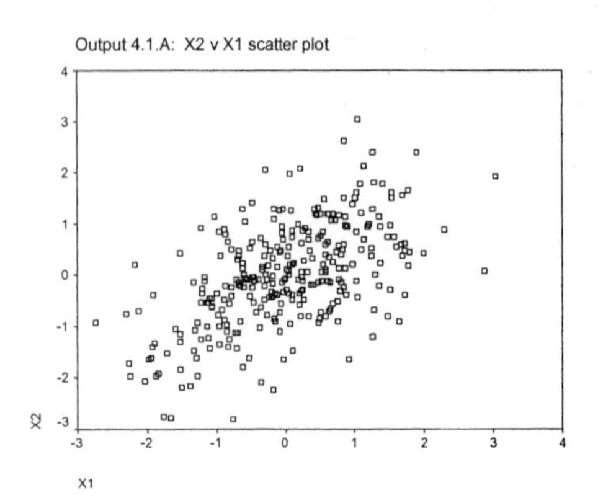

A scatter plot of the first two variables in the data set reveals a moderately strong correlation between X_1 and X_2 (the actual correlation of 0.562 is shown in the correlation matrix in SPSS Output 4.1.B). Scatter plots of X_3 versus X_1 and X_3 versus X_2 also show positive associations. The collinearity present in the data suggests that it should be possible to account for most of the information using fewer than three components.

Output 4.1.B *Results of principal component analysis of pca_example.sav*

Correlation Matrix

		X1	X2	X3
Correlation	X1	1.000	.562	.704
	X2	.562	1.000	.304
	X3	.704	.304	1.000

Eigenvalues are displayed for each of the three principal components as default output of the FACTOR command. The first eigenvalue ($\lambda_1 = 2.06$) suggests that principal component # 1 accounts for almost 70 percent of the variance in the original data, as shown in the column labeled "% of Variance".

Total Variance Explained

	Initial Eigenvalues			Extraction Sums of Squared Loadings		
Component	Total	% of Variance	Cumulative %	Total	% of Variance	Cumulative %
1	2.063	68.778	68.778	2.063	68.778	68.778
2	.706	23.518	92.296	.706	23.518	92.296
3	.231	7.704	100.000	.231	7.704	100.000

Extraction Method: Principal Component Analysis.

Principal component loadings are shown under the heading "Component Matrix." Loadings are the correlations between the original variables, labeled x1, x2 and x3, and the principal components labeled Component 1, 2, and 3. Finally, because we added the keyword FSCORE in the /PRINT subcommand, the "Component Score Coeffficient Matrix" is also displayed. These are the eigenvectors divided by the square root of the eigenvalue.

Component Matrix[a]

	Component		
	1	2	3
X1	.928	-7.98E-02	-.364
X2	.726	.670	.159
X3	.822	-.501	.271

Extraction Method: Principal Component Analysis.

a. 3 components extracted.

Component Score Coefficient Matrix

	Component		
	1	2	3
X1	.450	-.113	-1.575
X2	.352	.949	.688
X3	.398	-.710	1.171

Extraction Method: Principal Component Analysis.

Component Scores.

Output 4.1.C *Scatter plot of principal component loadings*

SPSS Output 4.1.C shows the scatter plot of the first two principal components from the illustrative example data. The plot shows that fac1_1 accounts for more variance than fac2_1 (together they account for more than 90 percent of the variance in the original data). The plot also shows that fac1_1 and fac2_1 are uncorrelated.

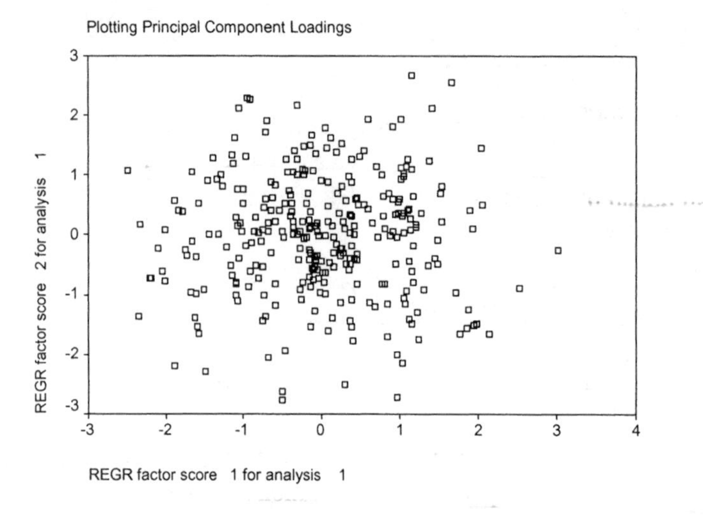

Output 4.1.D *Scatterplot Matrix for the variables fac1_1, fac2_1 and fac3_1*

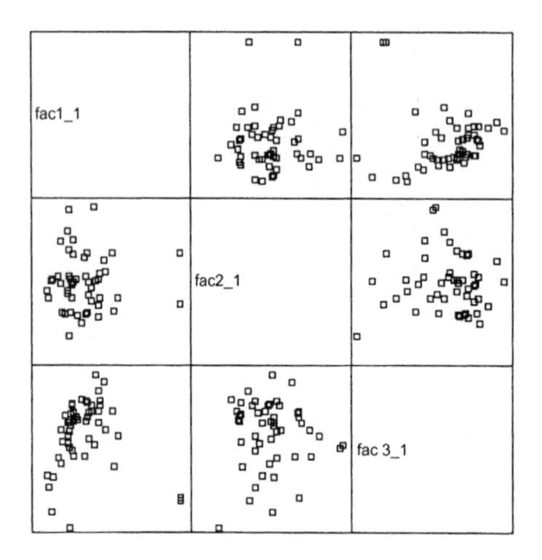

GRAPH /SCATTERPLOT(MATRIX) = [varlist] produces bivariate scatter plots for each pair of principal components specified by the variable list. Scatterplot matrices can also be interactively created from the **Graph** menu.

Example 4.2 *Gross State Product, Raw data*

Example 4.2 is a principal components analysis performed on the Gross State Product Raw data. The working data file is gsp_raw.sav, containing 13 dimensions of economic activity, expressed in millions of dollars. Variables are shown here renamed. The only output specified is a scree plot.

Example 4.2.SPS - SPSS Syntax Editor

File Edit View Analyze Graphs Utilities Run Window Help

```
*SPSS PROGRAM 4.2: Principal components analysis of GSP_RAW data.

FACTOR
 /VARIABLES agric mining constr durmfr nondmfr transp
commun utils whole retail fire  service govt
  /PLOT EIGEN.
```

SPSS Processor is ready

Output 4.2 *Scree plot for the GSP_RAW data*

The scree plot shows one large eigenvalue, consistent with the notion that a single principal component (reflecting the size of the state) captures more than 80 percent of the variation in these data.

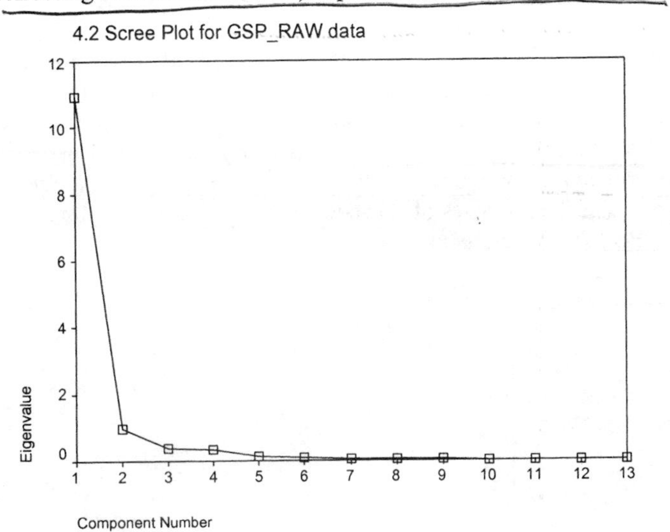

Example 4.3 *Gross State Product, Share data*

Example 4.3 runs a principal components analysis on the Gross State Product Share data, where each dimension of economic activity has been re-expressed as a proportion of the total economic activity in the state. The working data set is Gsp_Share.sav. Our goal here is to extract and examine the first few principal components.

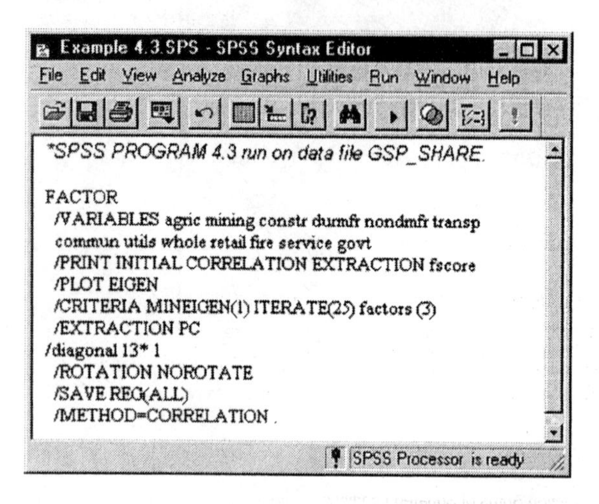

Output 4.3A *Scree plot for the GSP_SHARE data*

Compare the Share data scree plot to the previous Raw data scree plot (Output 4.2). By re-expressing the raw data as the share of economic activity in the state, we effectively remove the information on the total economic activity in the state. The remaining eigenvalues are much smaller than the first eigenvalue from the GSP_RAW data. Kaiser's rule, to accept components with Eigen value 1 or greater, suggests retaining the first five principal components, but the scree plot reveals a slight elbow between the third and fourth components. For the purposes of subsequent analysis, we retain three principal components using the FACTORS (3) subcommand.

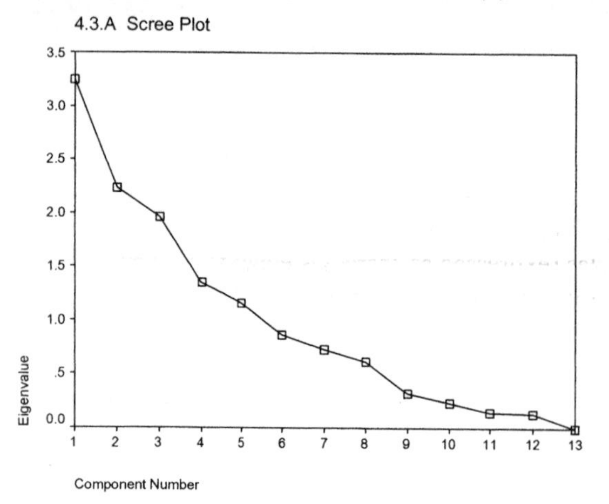

Output 4.3.B *Principal Components Analysis*

Communalities

	Initial	Extraction
agriculture	1.000	.349
mining	1.000	.846
construction	1.000	.480
durmfr	1.000	.701
nondmfr	1.000	.474
transp	1.000	.615
commun	1.000	.311
utils	1.000	.249
whole	1.000	.488
retail	1.000	.683
fiduciar	1.000	.819
service	1.000	.828
govt	1.000	.588

Extraction Method: Principal Component Analysis.

The "Communalities" table shows the proportion of variance in each of the original variables accounted for by the first three principal components. These are the values shown in Table 4.14 of the textbook. Note that the first three principal components capture more information in some areas of economic activity (e.g., fiduciary at 82 percent) than in others (e.g., utilities at 25 percent).

Total Variance Explained

	Initial Eigenvalues			Extraction Sums of Squared Loadings		
Component	Total	% of Variance	Cumulative %	Total	% of Variance	Cumulative %
1	3.236	24.889	24.889	3.236	24.889	24.889
2	2.236	17.204	42.092	2.236	17.204	42.092
3	1.960	15.076	57.168	1.960	15.076	57.168
4	1.360	10.464	67.632			
5	1.157	8.903	76.535			
6	.868	6.679	83.215			
7	.724	5.573	88.788			
8	.616	4.737	93.524			
9	.318	2.448	95.972			
10	.235	1.810	97.783			
11	.152	1.167	98.949			
12	.136	1.050	99.999			
13	6.855E-05	5.273E-04	100.000			

Extraction Method: Principal Component Analysis.

The output from the principal components analysis is shown in "Total Variance Explained". Note that the last eigenvalue is effectively zero (the small difference is attributable to round off error), which is due to the fact that the sum of the original variables (expressed as shares) is equal to 1. Thus, despite the fact that there are 13 areas of economic activity, there are only 12 principal components.

The "Component Matrix" reveals relatively high correlations between the first principal component and two areas of economic activity: mining (0.84) and transportation (0.75). Other variables with relatively high loadings on the first principal component are wholesale (–0.57), fiduciary (–0.65) and service (–0.68).

Component Matrix[a]

	Component		
	1	2	3
agriculture	.243	-1.12E-02	.539
mining	.845	-2.22E-03	-.364
construction	6.347E-02	.588	.360
durmfr	-.330	-.562	.526
nondmfr	-1.75E-02	-.686	4.997E-02
transp	.753	.220	8.901E-03
commun	-.273	.472	-.115
utils	.444	-.206	9.641E-02
whole	-.567	-4.23E-02	.406
retail	-.162	.390	.710
fiduciar	-.653	4.526E-02	-.625
service	-.683	.574	-.178
govt	.520	.551	.120

Extraction Method: Principal Component Analysis.

a. 3 components extracted.

A labeled plot of the first two principal components clearly shows Alaska (AK) and Wyoming (WY) as two outlying observations on the first principal component.

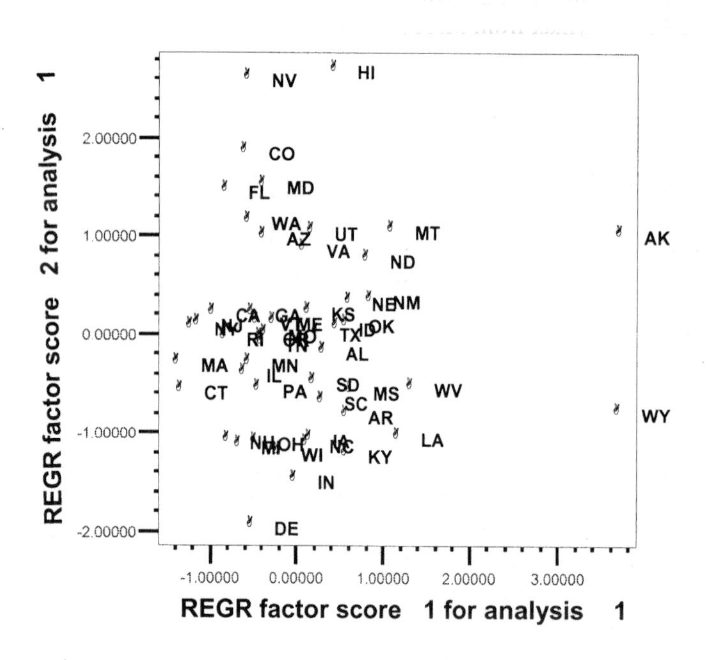

5 Exploratory Factor Analysis Using SPSS

Factor analysis examples in this chapter demonstrate estimation of appropriate number of factors; estimation of communalities and percent of variance accounted for by factors; plotting or rotating factor loadings to visualize relationships; and saving factor scores as variables for further analysis.

Example 5.1 *Factor Analysis of Psychological Testing data*

Define a correlation matrix of psychological test scores as the active data file. The command MATRIX DATA specifies how SPSS will read and define the matrix. The dataset is psychological testing data from Holzinger and Swineford, presented in Table 5.3 of Analyzing Multivariate Data. The command pair BEGIN DATA--END DATA signals the commencement and finish of the matrix values to be read by SPSS. Copy data for variables para, sent, word, add, and dots from the correlation matrix in file psych_tests.txt. Paste directly into the SPSS syntax file, between the BEGIN DATA and END DATA commands. Select the block of syntax from MATRIX DATA through END DATA and run. This creates a working matrix data file for Example 5.1.

Program 5.1matrix.SPS - SPSS Syntax Editor

File Edit View Analyze Graphs Utilities Run Window Help

```
*Principal factor analysis and iterated principal factor analysis.
*Define the input data from psych_tests.txt as an SPSS matrix dataset.
MATRIX DATA VARIABLES= para sent word add dots /N=145 /CONTENTS=corr.
BEGIN DATA
1
.722   1
.714   .685   1
.203   .246   .170   1
.095   .181   .113   .585   1
END DATA.
```

Example 5.1.A *Perform a first-round principal axis factor analysis with a single iteration and initial estimates of all communalities set to 0.5.*

The working dataset for this analysis is the psychological testing scores correlation matrix. Open the **Factor Analysis** dialog box with menu path **Analyze-Data Reduction-Factor**. Click on the **Extraction** button to specify the extraction method and number of factors within the. Select Principal Axis Factoring as **Method**. Specify the **Number of Factors** as two.

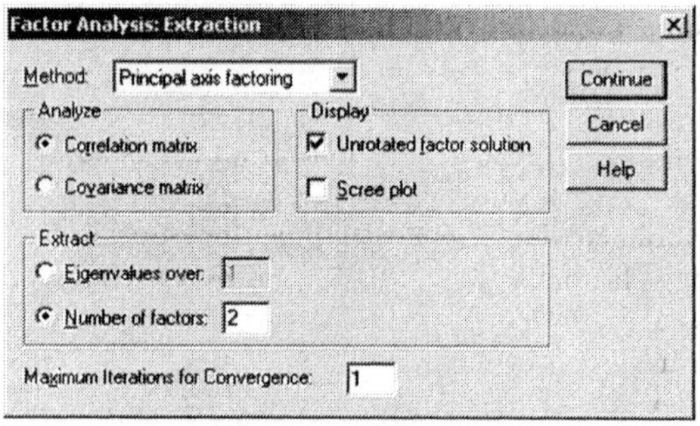

Factor Analysis: Extraction

Method: Principal axis factoring

Analyze: (•) Correlation matrix ( ) Covariance matrix

Display: [x] Unrotated factor solution [ ] Scree plot

Extract: ( ) Eigenvalues over: 1 (•) Number of factors: 2

Maximum Iterations for Convergence: 1

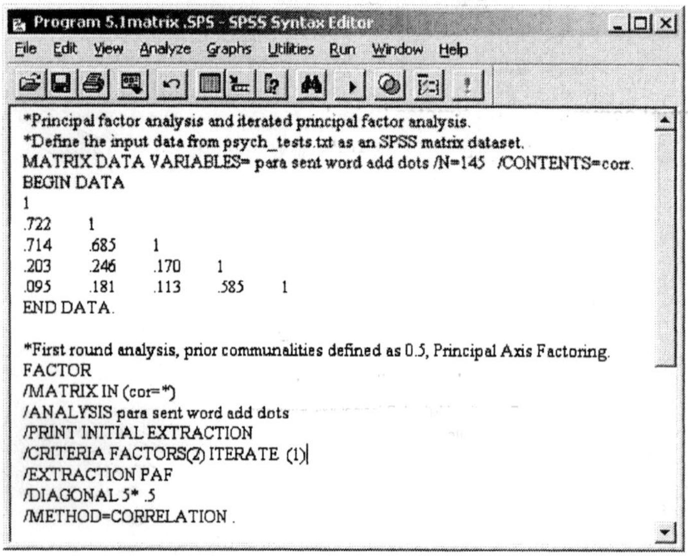

Continue Cancel Help

Program 5.1matrix.SPS - SPSS Syntax Editor

File Edit View Analyze Graphs Utilities Run Window Help

```
*Principal factor analysis and iterated principal factor analysis.
*Define the input data from psych_tests.txt as an SPSS matrix dataset.
MATRIX DATA VARIABLES= para sent word add dots /N=145 /CONTENTS=corr.
BEGIN DATA
1
.722    1
.714    .685    1
.203    .246    .170    1
.095    .181    .113    .585    1
END DATA.

*First round analysis, prior communalities defined as 0.5, Principal Axis Factoring.
FACTOR
/MATRIX IN (cor=*)
/ANALYSIS para sent word add dots
/PRINT INITIAL EXTRACTION
/CRITERIA FACTORS(2) ITERATE (1)
/EXTRACTION PAF
/DIAGONAL 5* .5
/METHOD=CORRELATION .
```

The /CRITERIA subcommand
Subcommand /CRITERIA FACTORS(2) iterate (1) specifies that two factors are extracted and that SPSS estimates are based on a single analysis iteration.

The /EXTRACTION paf subcommand specifies Principal Axis Factoring as the method of extraction.

Some syntax must be directly edited:

The subcommand /MATRIX IN is required for factor analysis of matrix data (/MATRIX IN substitutes for the /VARIABLES subcommand). /MATRIX IN (COR=*) runs analysis on the active data file, defining it as a correlation matrix.

Use the /DIAGONAL subcommand to specify initial communality estimates.

- /DIAGONAL 5* .5 defines priors as .5 for all 5 variables in the analysis.
- An alternative form: /DIAGONAL.5 .5 .5 .5 .5

The main difference between principal components analysis and the common factor model is the specification of communalities (the extent to which any particular variable is correlated to a factor). With the common factor model, we assume that the variance observed in each of our measures is attributable to two sources: a small number of common factors, and a single factor (often interpreted as measurement error) specific to a particular measure. The proportion of variance in the measure explained by the common factors is called the communality, and it is typically less than one. We can specify prior communalities (Example 5.1.A), or rely on the SPSS default estimation of communalities as equal to the squared multiple correlation (as in Example 5.1.B).

Output 5.1.A *Initial Communalities set at .5*

Note the difference between the initial communality estimates (which are all equal to 0.5 and sum to 2.5) and the final communality estimates (which are all greater than 0.5 and sum to 3).

Communalities

	Initial	Extraction
PARA	.500	.652
SENT	.500	.639
WORD	.500	.628
ADD	.500	.546
DOTS	.500	.544

Extraction Method: Principal Axis Factoring.

Factor Matrix[a]

	Factor	
	1	2
PARA	.772	-.235
SENT	.784	-.158
WORD	.756	-.237
ADD	.429	.602
DOTS	.348	.651

Extraction Method: Principal Axis Factoring.

a. Attempted to extract 2 factors. More than 1 iterations required. (Convergence=.152). Extraction was terminated.

Total Variance Explained

	Initial Eigenvalues			Extraction Sums of Squared Loadings		
Factor	Total	% of Variance	Cumulative %	Total	% of Variance	Cumulative %
1	2.587	51.749	51.749	2.087	41.749	41.749
2	1.422	28.434	80.184	.922	18.434	60.184
3	.415	8.304	88.488			
4	.311	6.222	94.710			
5	.265	5.290	100.000			

Extraction Method: Principal Axis Factoring.

Example 5.1.B *A follow-up analysis using squared multiple correlations as initial communality estimates and permitting multiple iterations.*

SPSS FACTOR sets priors equal to squared multiple correlations (SMC) by default, so syntax is identical to Example 5.1.A, without the /DIAGONAL subcommand.

Program 5.1matrix.SPS - SPSS Syntax Editor

File Edit View Analyze Graphs Utilities Run Window Help

```
*Principal factor analysis and iterated principal factor analysis.
*Define the input data from psych_tests.txt as an SPSS matrix dataset.
MATRIX DATA VARIABLES= para sent word add dots /N=145 /CONTENTS=corr.
BEGIN DATA
1
.722   1
.714   .685   1
.203   .246   .170   1
.095   .181   .113   .585   1
END DATA.

*First round analysis, prior communalities defined as 0.5, Principal Axis Factoring
FACTOR
/MATRIX IN (cor=*)
/ANALYSIS para sent word add dots
/PRINT INITIAL EXTRACTION
/CRITERIA FACTORS(2) ITERATE (1)
/EXTRACTION PAF
/DIAGONAL 5* .5
/METHOD=CORRELATION .

*Priors estimated by squared multiple correlation, Unweighted Least Squares factoring
FACTOR
/MATRIX IN (cor=*)
/ANALYSIS para sent word add dots
/PRINT INITIAL EXTRACTION
/CRITERIA FACTORS(2) ITERATE (15)
/EXTRACTION ULS
/METHOD=CORRELATION .
```

The initial communality estimates provided by the squared multiple correlations differ noticeably across tests: the first three (*PARA, SENT, WORD*) are all in the vicinity of 0.60, while the last two (*ADD, DOTS*) are closer to 0.35. This suggests (at least initially) differences in reliability across measures. Through 7 iterations of the procedure, the estimated communalities all increase. At convergence, they range from 0.48 to 0.76 and sum to 3.3. The two common factors together account for 3.3/5 = 66 percent of the variance in the five psychological tests. The remaining variation is accounted for by the specific factors associated with each test.

Communalities

	Initial	Extraction
PARA	.616	.757
SENT	.591	.701
WORD	.570	.673
ADD	.367	.475
DOTS	.349	.746

Extraction Method: Unweighted Least Squares.

Factor Matrix[a]

	Factor	
	1	2
PARA	.834	-.247
SENT	.825	-.144
WORD	.788	-.228
ADD	.396	.564
DOTS	.358	.787

Extraction Method: Unweighted Least Squares.
a. 2 factors extracted. 7 iterations required.

Total Variance Explained

Factor	Initial Eigenvalues			Extraction Sums of Squared Loadings		
	Total	% of Variance	Cumulative %	Total	% of Variance	Cumulative %
1	2.587	51.749	51.749	2.283	45.650	45.650
2	1.422	28.434	80.184	1.070	21.404	67.054
3	.415	8.304	88.488			
4	.311	6.222	94.710			
5	.265	5.290	100.000			

Extraction Method: Unweighted Least Squares.

Example 5.2 *Rotation of Factor Scores*

Example 5.2 demonstrates orthogonal (VARIMAX) rotation of hypothetical data on pain relievers. The working data set is pain_relief.sav. Rename variables col1, co2, col3, col4, col5 col6 to stomach effects quickly stops awake limited. A bivariate scatterplot of factor loadings portrays relationships between the common factors and analysis variables.

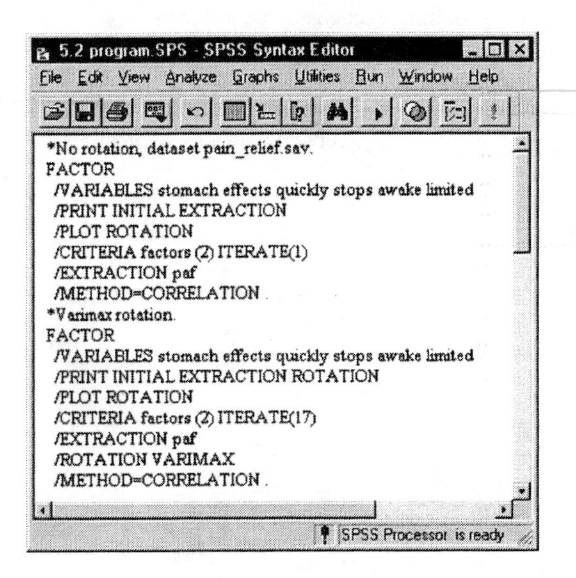

Output 5.2.A *Unrotated Factor Loadings*

All of the unrotated factor loadings (called the factor matrix) are greater than 0.40 in absolute value, which makes it very hard to interpret the underlying factors. The final communality estimate for each variable is greater than 50 percent. The two factors together account for (1.92 + 1.56) / 6 = 58 percent of the variance in the original data,

Communalities

	Initial	Extraction
STOMACH	.453	.539
EFFECTS	.509	.600
QUICKLY	.517	.606
STOPS	.499	.588
AWAKE	.487	.582
LIMITED	.479	.568

Extraction Method: Principal Axis Factoring.

Factor Matrix[a]

	Factor	
	1	2
STOMACH	.579	-.452
EFFECTS	.522	-.572
QUICKLY	.645	.436
STOPS	.542	.542
AWAKE	-.476	.596
LIMITED	-.613	-.439

Extraction Method: Principal Axis Factoring.

a. Attempted to extract 2 factors. More than 1 iterations required. (Convergence=9.456E-02). Extraction was terminated.

Total Variance Explained

	Initial Eigenvalues			Extraction Sums of Squared Loadings		
Factor	Total	% of Variance	Cumulative %	Total	% of Variance	Cumulative %
1	2.431	40.512	40.512	1.921	32.022	32.022
2	2.070	34.498	75.010	1.562	26.030	58.052
3	.439	7.319	82.329			
4	.387	6.450	88.778			
5	.380	6.340	95.118			
6	.293	4.882	100.000			

Extraction Method: Principal Axis Factoring.

Output 5.2.B *Varimax Rotated Factor Loadings*

Output 5.2.B shows the Varimax Rotated Factor Matrix rotated factor loadings for the pain reliever data. As explained in the textbook, Varimax rotation tries to achieve simple structure by focusing on the columns of the factor loadings matrix (trying to drive the squared loadings toward 0 or 1 by maximizing the column variance). The factor pattern is now very different. Each variable loads primarily on one of the two factors. In each case, primary loading exceeds 0.70, and the cross-loading is never greater than 0.15. This suggests a clear interpretation: Factor 1 is positively associated with "works quickly" and "stops the pain" and negatively associated with "limited relief" suggesting that the factor is interpretable as effectiveness. Factor 2 is positively associated with "no upset stomach" and "no side effects" and negatively associated with "keeps me awake" suggesting an interpretation of gentleness. Final communality estimates are similar from the unrotated to the rotated solution, demonstrating that the communalities are unaffected by orthogonal rotation. In Total Variance Explained note that the variance accounted for by each factor has changed from the unrotated solution: Factor 1 now accounts for 2.0 and Factor 2 accounts for 1.7. However, the total proportion of variance accounted for shows little change from the unrotated to the rotated solution (from 58 to 62 percent).

Communalities

	Initial	Extraction
STOMACH	.453	.569
EFFECTS	.509	.652
QUICKLY	.517	.652
STOPS	.499	.634
AWAKE	.487	.637
LIMITED	.479	.609

Extraction Method: Principal Axis Factoring.

Rotated Factor Matrix [a]

	Factor	
	1	2
STOMACH	.141	.741
EFFECTS	1.440E-02	.808
QUICKLY	.801	9.787E-02
STOPS	.794	-5.525E-02
AWAKE	3.900E-02	-.797
LIMITED	-.777	-7.384E-02

Extraction Method: Principal Axis Factoring.
Rotation Method: Varimax with Kaiser Normalization.
a. Rotation converged in 3 iterations.

Total Variance Explained

	Initial Eigenvalues			Extraction Sums of Squared Loadings			Rotation Sums of Squared Loadings		
Factor	Total	% of Variance	Cumulative %	Total	% of Variance	Cumulative %	Total	% of Variance	Cumulative %
1	2.431	40.512	40.512	2.055	34.250	34.250	1.899	31.647	31.647
2	2.070	34.498	75.010	1.699	28.315	62.566	1.855	30.918	62.566
3	.439	7.319	82.329						
4	.387	6.450	88.778						
5	.380	6.340	95.118						
6	.293	4.882	100.000						

Extraction Method: Principal Axis Factoring.

Output 5.2.C *Plot of the rotated factor loadings*

Each factor appears associated with a subset of the original variables. Factor 1 loads on Quickly and Stops. Factor 2 loads on Effects and Stomach.

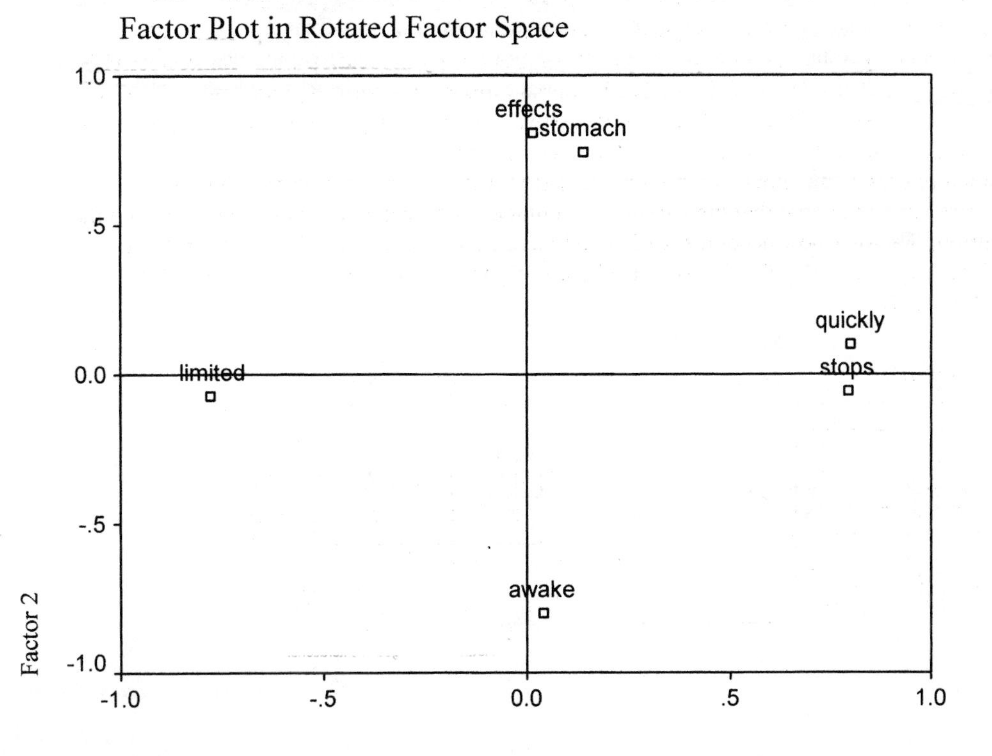

Example 5.3 *Analysis of RTE Cereal Data*

Example 5.3 demonstrates how to use common factor analysis to reduce a set of 25 variables to four common factors. This is a common factor analysis (with varimax rotation) The working dataset is rte_cereal.sav, rename variables according to the list below or see the textbook discussion of Lattin and Roberts' multiattribute model of choice (section 5.3.1). Recall that 116 respondents provided 235 evaluations of 12 brands on 25 different attributes. Lattin and Roberts used factor analysis to reduce the dimensionality of the space from 25 attributes to four common factors. Example 5.3.A demonstrates a principal components analysis and scree plot. These suggest we are justified extracting four common factors. Example 5.3.B demonstrates extract four common factors by Principal Axis factoring.

5.3.SPS - SPSS Syntax Editor

```
*Use Principal Components and scree plot to suggest number of factors.
FACTOR
 /VARIABLES filling natural fibre sweet easy salt satis energy fun kids soggy econ health
 family calories plain crisp regular sugar fruit process quality treat boring nutri
 /PLOT EIGEN /EXTRACTION PC
 /CRITERIA MINEIGEN(1) ITERATE(10) /METHOD=CORRELATION .

*Common factor model with 4 factors.
FACTOR
 /VARIABLES filling natural fibre sweet easy salt satis energy fun kids soggy econ health
 family calories plain crisp regular sugar fruit process quality treat boring nutri
 /PRINT INITIAL EXTRACTION ROTATION FSCORE
 /CRITERIA factor (4) iterate (25)
 /EXTRACTION paf
 /ROTATION varimax
 /METHOD=CORRELATION .
```

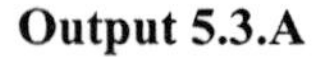

Output 5.3.A

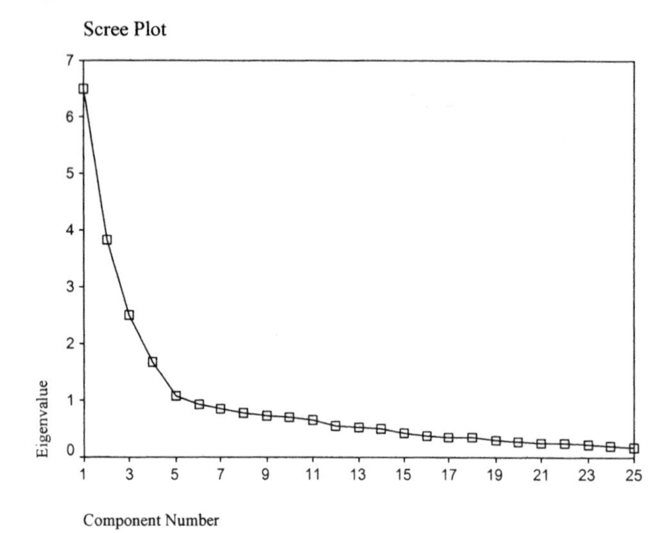

Output 5.3.B shows the factor loadings matrix for the RTE cereal data after varimax rotation. Each loading is interpretable as the correlation between the original variable (row) and the underlying common factor (column). Thus, the correlation between the attribute "filling" and Factor 1 is 0.706. Examining the list of attributes highly correlated with a factor (either positively or negatively) helps to provide an interpretation for that factor. For example, the attributes strongly correlated with Factor 1 (all above 0.60) are: filling, natural, fibre, satisfying, energy, health, regular, quality, nutritious. These attributes are all consistent with the interpretation "beneficial" or "good for you." Note that the structure is relatively simple (i.e., the incidence of high cross-loadings of attributes on factors is low), which facilitates interpretation.

Rotated Factor Matrix [a]

	Factor			
	1	2	3	4
FILLING	.706	8.815E-02	.199	.151
NATURAL	.753	-.209	5.506E-02	3.663E-02
FIBRE	.821	-.116	-.121	2.113E-02
SWEET	6.886E-02	.702	7.396E-02	.347
EASY	.238	6.338E-02	.325	6.443E-02
SALT	-9.190E-02	.686	1.587E-02	-8.345E-02
SATIS	.626	7.681E-02	.424	.171
ENERGY	.660	7.822E-02	.192	.210
FUN	.163	.176	.417	.478
KIDS	-2.464E-02	3.438E-02	.850	1.046E-02
SOGGY	3.314E-02	1.440E-02	9.427E-02	-.481
ECON	6.870E-02	-.281	.415	-.229
HEALTH	.829	-.288	5.147E-02	4.601E-02
FAMILY	6.208E-02	-5.509E-02	.761	9.011E-02
CALORIES	-.114	.627	-7.444E-03	.120
PLAIN	-.147	-6.201E-02	6.832E-02	-.657
CRISP	7.374E-02	.146	.373	.436
REGULAR	.613	-9.997E-02	-2.740E-02	8.899E-02
SUGAR	-.184	.817	-5.253E-02	.165
FRUIT	.376	.187	-.267	.443
PROCESS	-.236	.374	2.619E-02	-.126
QUALITY	.647	-.245	.204	.170
TREAT	.245	.233	.336	.602
BORING	-.165	6.685E-02	-.225	-.505
NUTRI	.831	-.177	5.116E-02	5.588E-02

Extraction Method: Principal Axis Factoring.
Rotation Method: Varimax with Kaiser Normalization.

a. Rotation converged in 6 iterations.

Output 5.3.C shows the final communality estimates. Some of the values are quite low (e.g., "easy" with 0.171) which can be a source of concern. Recall from the text that this may be due either to an attribute with low reliability (i.e., an inherently noisy measure) or to the fact that the attribute is the only measure of an independent underlying factor.

Communalities

	Initial	Extraction
FILLING	.633	.569
NATURAL	.636	.615
FIBRE	.703	.703
SWEET	.612	.623
EASY	.230	.171
SALT	.448	.486
SATIS	.615	.606
ENERGY	.571	.522
FUN	.483	.461
KIDS	.636	.725
SOGGY	.314	.241
ECON	.372	.308
HEALTH	.750	.774
FAMILY	.599	.594
CALORIES	.417	.420
PLAIN	.408	.461
CRISP	.410	.356
REGULAR	.521	.395
SUGAR	.664	.731
FRUIT	.488	.444
PROCESS	.297	.212
QUALITY	.614	.549
TREAT	.561	.590
BORING	.334	.337
NUTRI	.711	.728

Extraction Method: Principal Axis Factoring.

Example 5.4 *Save and plot average factor scores*

For this example create syntax identical to 5.3.B, with an additional subcommand to save factor scores. Select the **Scores** button within the **Factor Analysis** dialog box.

Factor Analysis: Factor Scores

Save as variables
Method
Regression
Bartlett
Anderson-Rubin
Display factor score coefficient matrix

Continue
Cancel
Help

Saved factor scores are standardized (i.e., mean 0 and variance 1) and added to the working data file with the default variable names fac1_1, fac1_2, and so on.

Lattin and Roberts used the average factor score to represent the position of the brand for each individual. Follow menu path **Analyze-Compare Means-Means** to calculate average factor scores for each brand of cereal with the MEANS command. Finally the GRAPH command labels the means and plots the average for each brand in the factor space. Pasted syntax appears as follows:

```
*Common factor model with 4 factors, Varimax rotation, Factor scores saved.
FACTOR
  /VARIABLES filling natural fibre sweet easy salt satis energy fun kids
  soggy econ health family calories plain crisp regular sugar fruit process
  quality treat boring nutri
  /PRINT INITIAL EXTRACTION ROTATION FSCORE
  /CRITERIA FACTORS(4) ITERATE(25)
  /EXTRACTION PAF
  /CRITERIA ITERATE(25)
  /ROTATION VARIMAX
  /SAVE REG(ALL)
  /METHOD=CORRELATION .

*Calculate average factor scores for 235 respondents.
MEANS
  TABLES=fac1_1 fac2_1 fac3_1 fac4_1 BY brand
  /CELLS MEAN.

*Copy average factor scores into Excel file, then open as SPSS working data file.
*Define variables Brand, Avfac1_1, Avfac2_1, etc.

*Bivariate scatter plot of Average Factor score 1 and Average Factor score 2.
GRAPH
  /SCATTERPLOT(BIVAR)=Avfac2_1 WITH Avfac1_1 BY brand.
```

Output 5.4.A: *RTE cereal data factor score coefficients and average factor scores by brand*

Factor Score Coefficient Matrix

	Factor			
	1	2	3	4
FILLING	.123	.072	.016	-.034
NATURAL	.134	-.010	-.020	-.039
FIBRE	.233	.067	-.126	-.078
SWEET	.045	.208	-.015	.082
EASY	.003	.014	.056	-.004
SALT	.028	.208	.016	-.132
SATIS	.120	.075	.149	-.034
ENERGY	.080	.076	.030	.015
FUN	.007	.000	.088	.120
KIDS	-.056	.021	.471	-.105
SOGGY	.024	.050	.028	-.161
ECON	-.001	-.035	.093	-.070
HEALTH	.249	-.071	.000	-.069
FAMILY	-.015	-.046	.223	.003
CALORIES	.030	.150	-.008	-.022
PLAIN	.037	.083	.083	-.294
CRISP	-.034	-.016	.071	.136
REGULAR	.032	-.036	-.017	.025
SUGAR	.054	.438	-.029	-.048
FRUIT	.028	.029	-.113	.151
PROCESS	.007	.070	.012	-.065
QUALITY	.037	-.053	.057	.036
TREAT	-.031	.030	.067	.273
BORING	.035	.071	-.016	-.180
NUTRI	.225	.005	-.032	-.076

Extraction Method: Principal Axis Factoring.
Rotation Method: Varimax with Kaiser Normalization.
Factor Scores Method: Regression.

Report

Mean

BRAND	REGR factor score 1 for analysis 1	REGR factor score 2 for analysis 1	REGR factor score 3 for analysis 1	REGR factor score 4 for analysis 1
1	.3482327	-.3190490	-.8860911	-.3756282
3	.5175271	.4988810	-.2339592	.6582945
13	-1.3497947E-02	.2275109	-.4197648	.4752301
14	-.5536089	.1171450	.5741343	2.417504E-02
15	.5457839	.2328251	-1.0089811	.6022963
16	-.4246676	.8090920	.5480202	.2854730
17	.6226729	.6831375	-.4104264	.5754575
19	-1.1648888	-.4284380	.6050718	6.262920E-02
21	-.3057843	-.2138710	.1544638	-5.4967988E-02
23	.6766800	-.3028083	-.3239443	.8565512
24	.3647640	-.5920290	.2183082	-.9316240
25	.3263499	-.4120884	-7.1154699E-02	-.8837226
Total	2.959872E-17	6.938894E-17	-9.8689503E-17	-2.7321895E-17

Output 5.4.B *Plot of average factors scores: Factor 2 versus Factor 1*

To make it easier to visualize their relative positions, the average scores for Factors 1 and 2 are plotted. The plot demonstrates that Purina, Cerola, Komplete, three brands of muesli, are positioned similarly with respect to the first two factors.

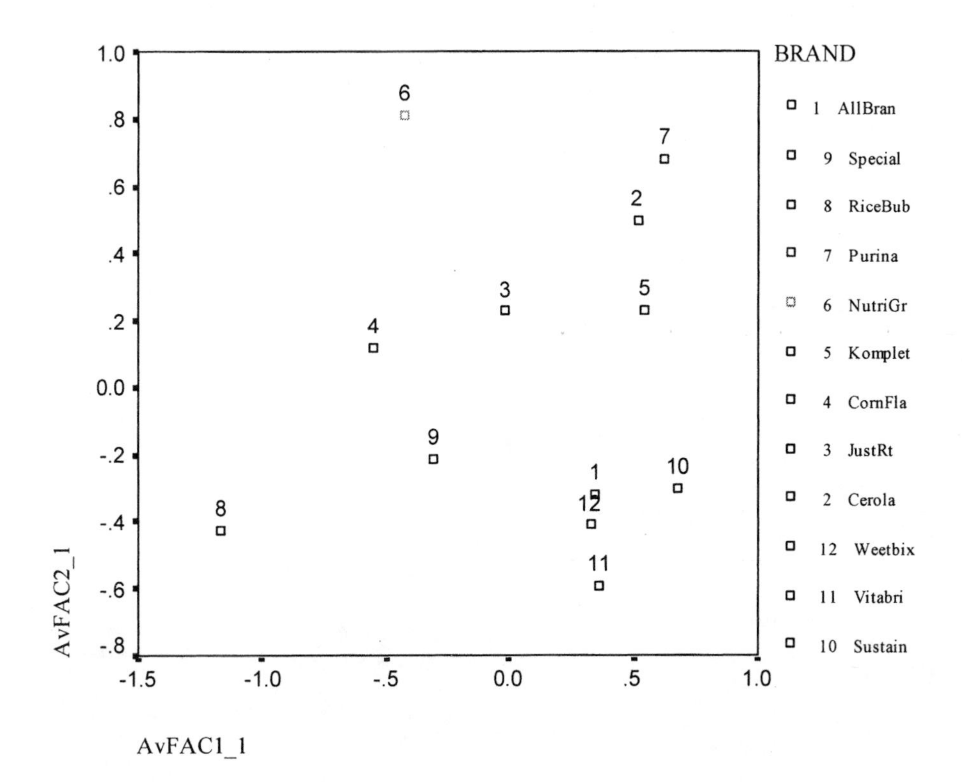

6 Confirmatory Factor Analysis Using Amos

Confirmatory factor analysis examples in this chapter employ Amos. Amos is an analysis of covariance and structural equation modeling software with graphical programming features. Examples presented in Guide Chapters 6 and 10 employ the stand-alone student version of Amos. The student version of Amos can be downloaded free from the Smallwaters Corporation homepage at http://www.smallwaters.com/amos/student.html. Amos is SPSS-licensed and can be installed as a component of Base SPSS.

Example 6.1 *Defining a correlation matrix data file for Amos analysis*

This syntax example converts a working rectangular SPSS data file into correlation matrix format, to be used in this chapter's examples. The Amos File menu imports numerous other data formats.

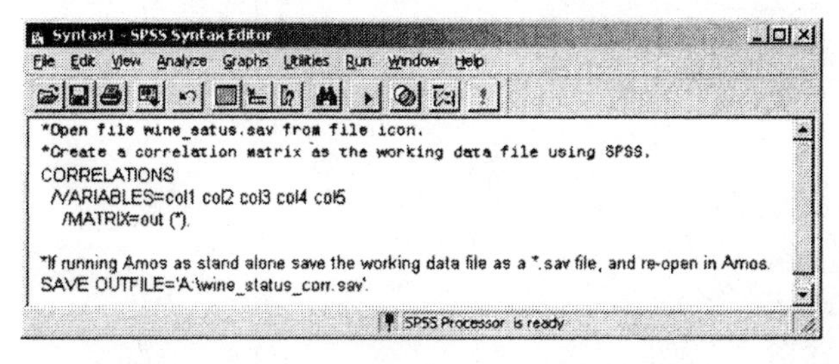

```
*Open file wine_satus.sav from file icon.
*Create a correlation matrix as the working data file using SPSS.
CORRELATIONS
 /VARIABLES=col1 col2 col3 col4 col5
  /MATRIX=out (*).

*If running Amos as stand alone save the working data file as a *.sav file, and re-open in Amos.
SAVE OUTFILE='A:\wine_status_corr.sav'.
```

The wine_status matrix file contains means, standard deviations, number of observations and correlation coefficients. The automatically created variable rowtype_ identifies and distinguishes these data types. Variables col1 through col5 represent expert ratings.

wine_status_corr.sav - SPSS Data Editor

	rowtype_	varname_	col1	col2	col3	col4	col5
1	MEAN		5.050847	3.813559	3.118644	4.508475	3.322034
2	STDDEV		1.419370	1.978109	1.390668	1.735759	1.906800
3	N	COL1	59.00000	59.00000	59.00000	59.00000	59.00000
4	N	COL2	59.00000	59.00000	59.00000	59.00000	59.00000
5	N	COL3	59.00000	59.00000	59.00000	59.00000	59.00000
6	N	COL4	59.00000	59.00000	59.00000	59.00000	59.00000
7	N	COL5	59.00000	59.00000	59.00000	59.00000	59.00000
8	CORR	COL1	1.000000	.7648959	.6782054	.6751495	.6818559
9	CORR	COL2	.7648959	1.000000	.7352164	.6256433	.7658457
10	CORR	COL3	.6782054	.7352164	1.000000	.6317007	.7135613
11	CORR	COL4	.6751495	.6256433	.6317007	1.000000	.5174844
12	CORR	COL5	.6818559	.7658457	.7135613	.5174844	1.000000

Example 6.2 *Amos Graphics*

Click on the Amos Graphics icon to open an Amos graphics document. The Amos graphic document resembles a drawing pad with extensive menus and toolbars, so that the user can visually diagram the relationship being examined and directly indicate within the diagram how Amos will interpret the model.

Graphically specify a model using the toolbar buttons or pull-down menu commands. Drag the toolbar to any part of the computer screen, or close and use the dropdown menus. Toolbar icons toggle on and off.

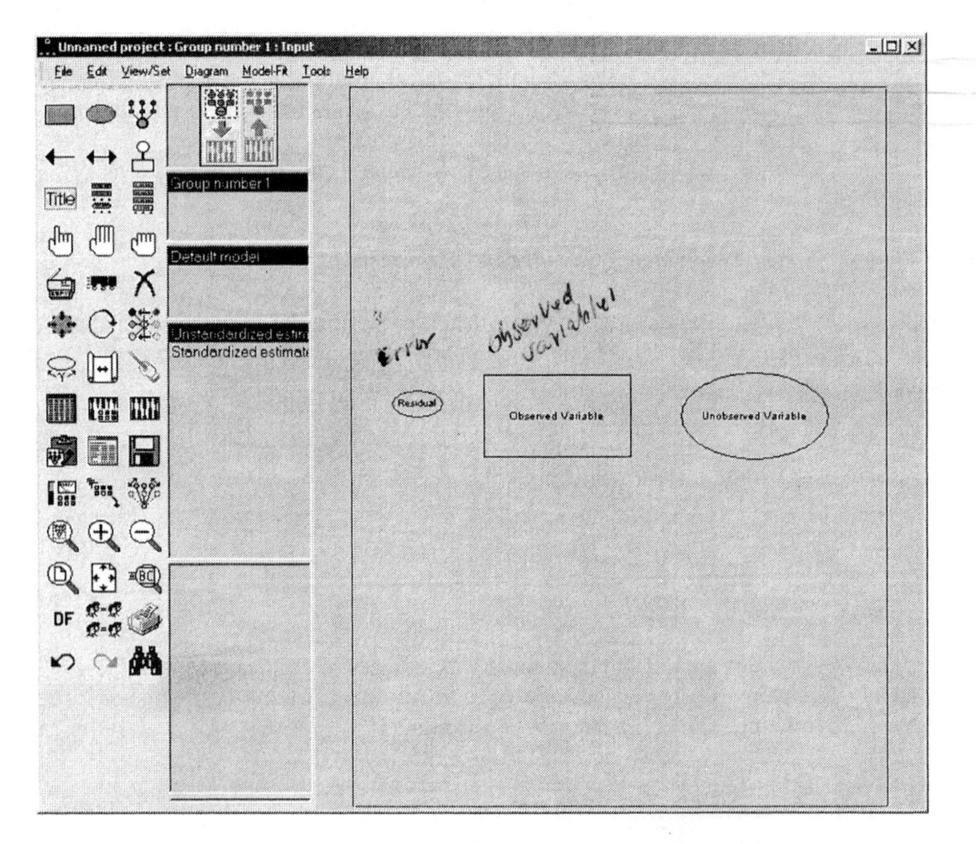

Observed variables are represented by rectangles. Click the menu **Diagram...Draw Observed**, keystroke **F3**, or highlight the **Draw Observed Variables** icon in the upper left corner of the Amos Toolbar. While the **Draw Observed Variables** icon is highlighted a new rectangle appears each time the left mouse button is depressed and the mouse is moved. Unobserved variables are similarly added from the menu item **Diagram...Draw Unobserved**, keystroke **F4**, or by highlighting the **Draw Unobserved Variables** icon in the upper left corner. Unobserved variables are represented by ellipses. Do not be concerned initially about neatness. The Amos toolbar contains 'tidy up' tools.

Example 6.3 *Graphically specify a Confirmatory One-Factor Model of Wine Status*

This analysis is designed to assess whether the wine status data are consistent with a single-factor model. Click on the Amos Graphics icon to open an Amos graphics document, and then confirm that Amos can see the wine_status.sav file as working file. In **File** menu select **Data Files** and File **Name**. Select wine_status.sav in the **Data Files** window and click OK.

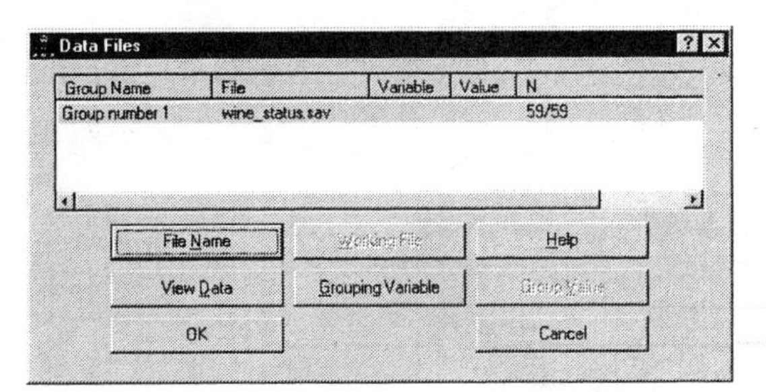

Example 6.3.A *Graphically specify the model*

As in Example 6.2.graphically specify the model using the toolbar buttons or pull-down menu commands. The five wine expert ratings are represented by rectangles. Click the menu **Diagram...Draw Observed**, keystroke **F3**, or highlight the **Draw Observed Variables** icon in the upper left corner of the Amos Toolbar. While the **Draw Observed Variables** icon is highlighted a new rectangle appears each time the left mouse button is depressed and the mouse is moved. Unobserved variables are similarly added from the menu item **Diagram...Draw Unobserved**, keystroke **F4**, or by highlighting the **Draw Unobserved Variables** icon in the upper left corner. The oval on the right represents a common factor of the five ratings. The five ovals on the left represent error terms. Any endogenous variable is required to have an associated oval representing the unobserved residual value.

Example 6.3.B *Naming Variables*

Name variables in the **Object Properties** dialog box. Open the **Object Properties** dialog box by double clicking on each variable shape in the model. Click on the **Text** tab and fill in the variable label box.

Object Properties
Colors | Text | Parameters | Format | V
Font size: 18
Font style: Regular
Variable name: col1
Variable label: col1
Set Default
Undo

Amos inserts the dataset variable name for observed variables where no label is specified.

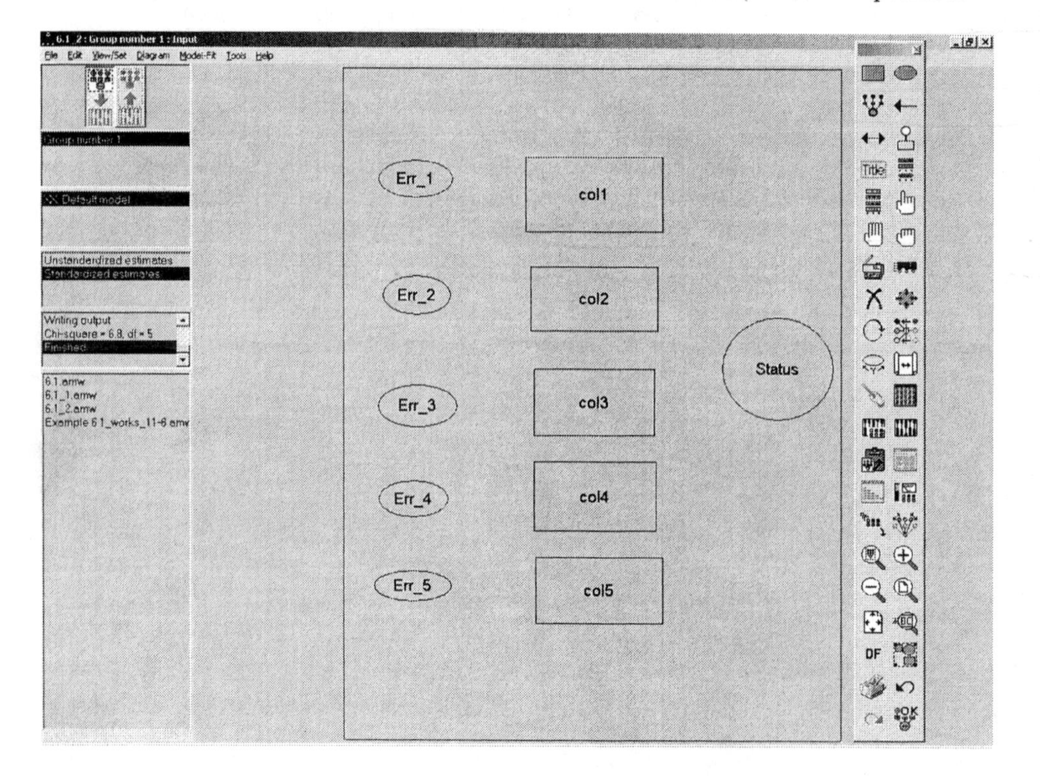

For more information about formatting diagrams refer to *Amos Graphics Online Help*. Documents on how to draw models can also be downloaded from the Smallwaters Corporation homepage, http://www.smallwaters.com.

Example 6.3.C *Specify relationships among variables: Paths and Parameter Values*

Click on the **Arrow** button in the toolbar or use Amos editing tools to draw path arrows. Arrows are automatically redrawn as path elements are moved about. To specify regression weights click the **Parameters** tab in the **Object Properties** dialog box, or double click on a path arrow.

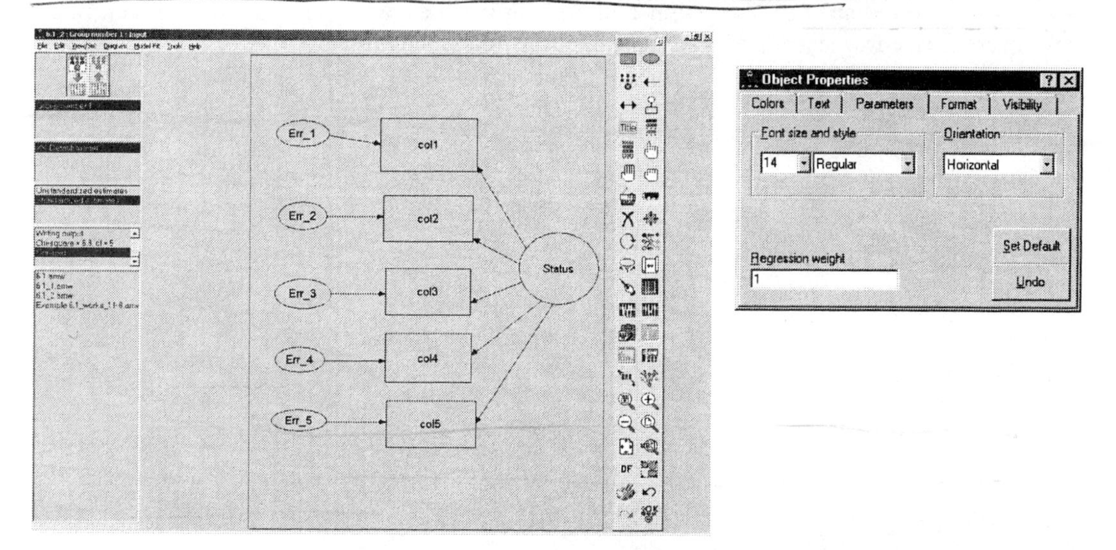

Some arbitrary value must be assigned to each error term path in order to set a measurement scale which Amos will rely on to estimate the remaining model coefficients. Each of the error terms in the following diagram is therefore assigned a path coefficient of 1. In order to completely identify the model one path between the common factor and an observed variable (col1) is also fixed at one.

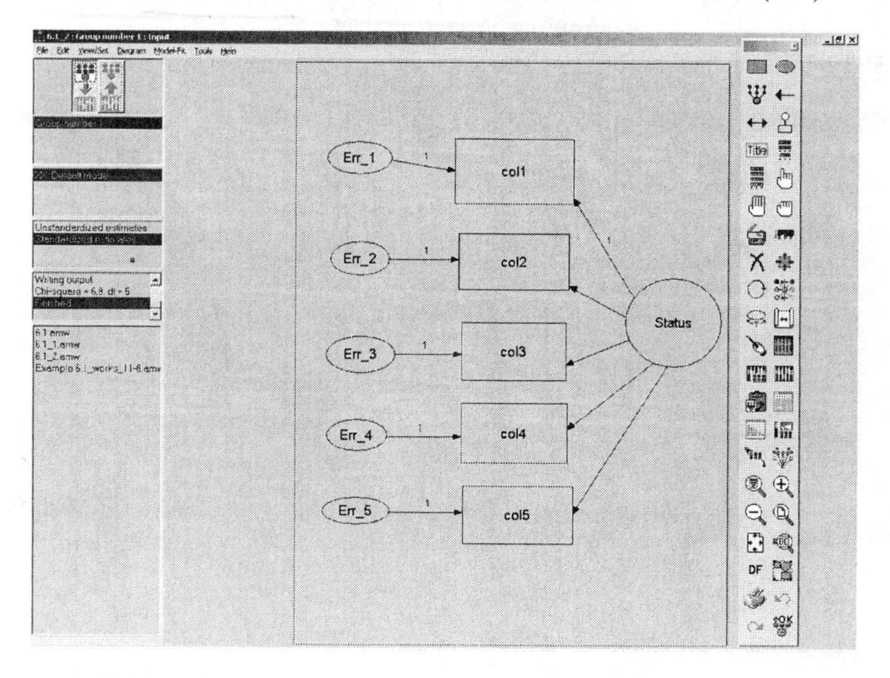

Example 6.4 *Analyzing an Amos model*

To specify terms for parameter estimation select the **View/Set** menu, **Analysis Properties** dialog box and **Estimation** tab. Default in AMOS is maximum likelihood, selected by clicking the first radio button of the **Estimation** tab. Other estimation methods include unweighted least squares, generalized least squares, scale-free least squares and asymptotically distribution-free.

Analysis Properties

Permutations | Random # | Title

Output formatting | Output | Bootstrap

Estimation | Numerical | Bias

Discrepancy

- (•) Maximum likelihood
- ( ) Generalized least squares
- ( ) Unweighted least squares
- ( ) Scale-free least squares
- ( ) Asymptotically distribution-free

- [ ] Estimate means and intercepts
- [ ] Emulisrel6
- [ ] Chicorrect

For the purpose of computing fit measures with incomplete data:

- (•) Fit the saturated and independence models
- ( ) Fit the saturated model only
- ( ) Fit neither model

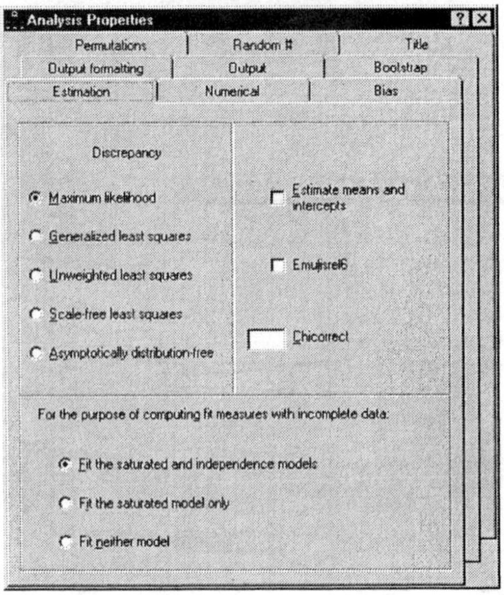

Specify AMOS output in the **Output** tab of the **Analysis Properties** dialog box. Select the Minimization History, Standardized Estimates and Squared Multiple Correlations check boxes.

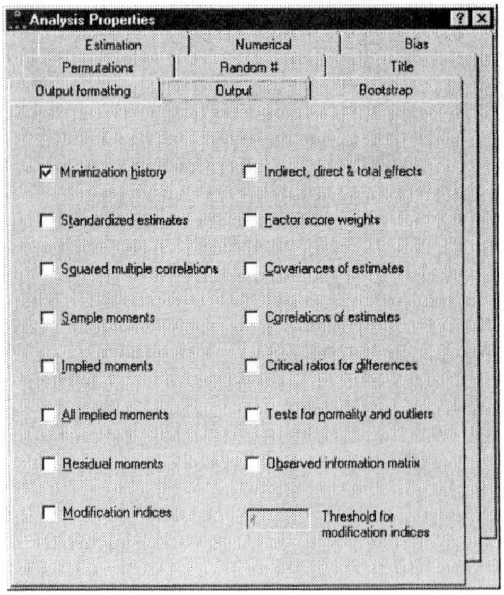

Analysis Properties

Estimation | Numerical | Bias

Permutations | Random # | Title

Output formatting | Output | Bootstrap

- [x] Minimization history
- [ ] Standardized estimates
- [ ] Squared multiple correlations
- [ ] Sample moments
- [ ] Implied moments
- [ ] All implied moments
- [ ] Residual moments
- [ ] Modification indices
- [ ] Indirect, direct & total effects
- [ ] Factor score weights
- [ ] Covariances of estimates
- [ ] Correlations of estimates
- [ ] Critical ratios for differences
- [ ] Tests for normality and outliers
- [ ] Observed information matrix
- Threshold for modification indices: 4

Output 6.4

Click on the **Calculate Estimates** icon in the toolbar to initiate calculations. Before initiating calculations Amos presents a **Save** As dialog box, which requires that input specifications be saved to a separate file (.amw). Amos output can be viewed as graphical object, table or as a single text document.

Output 6.4.A *Graphical Output*

View parameter estimates in the path diagram by clicking the **View the output path diagram** icon in the upper left corner of the screen. Then in the separate menu box below, select standardized or unstandardized estimates.

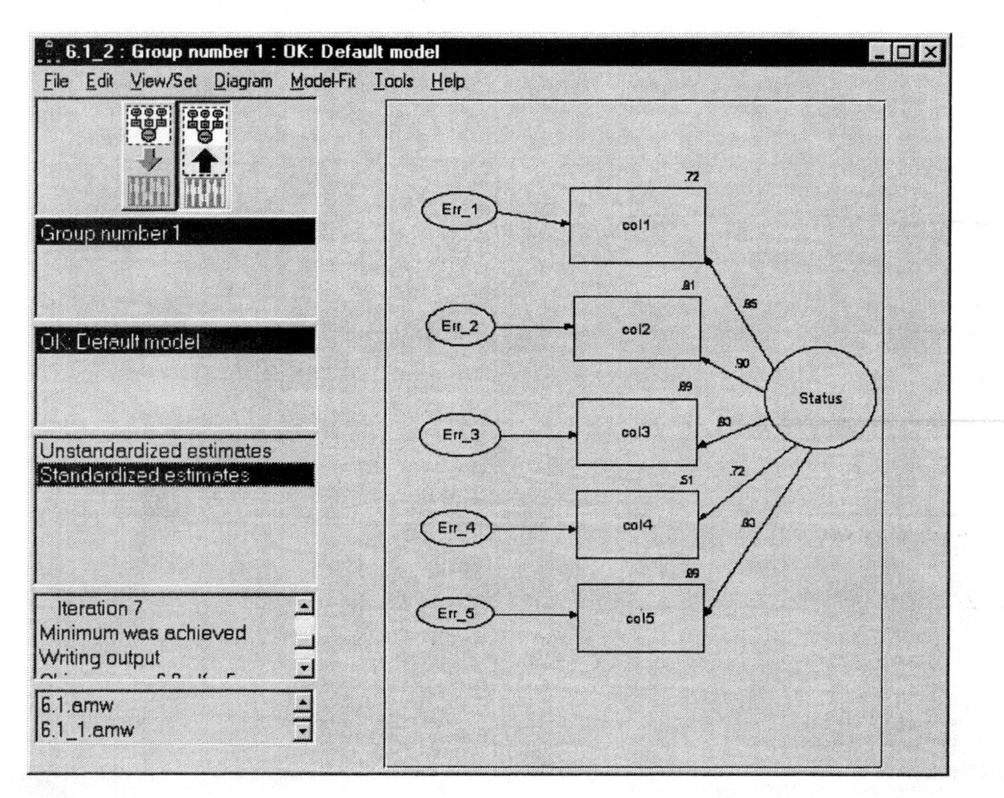

Output 6.4.B *Table Output*

To view table output click on the **View Spreadsheets** icon in the toolbar. Output can be viewed as tabs in the Amos Table Viewer output file (.amp extension) and can be directly copy-pasted to Excel.

C:\AMOS practice SPSS WB 6

File Edit Format Help

100%

Title
Variable Summary
Parameter Summary
Notes for Group
Notes for Model
Normality
Outliers
Estimates

Summary of Parameters

	Weights	Covariances	Variances	Means	Intercepts	Total
Fixed	5	0	1	0	0	6
Labeled	0	0	0	0	0	0
Unlabeled	5	0	5	0	0	10
Total	10	0	6	0	0	16

In the Summary of Parameters table the Fixed Weights are the error term parameter values, as well as the path between the Status factor and observed variable Col1, which was set at a value of 1. Fit measures are displayed in vertical or horizontal tables. The Fit Measure Chi-Square Probability level of .240 suggests a good model fit. Maximum Likelihood Estimates are estimated path coefficients.

C:\AMOS practice SPSS WB 6

File Edit Format Help

100%

Title
Variable Summary
Parameter Summary
Notes for Group
Notes for Model
Normality
Outliers
Minimization History
Estimates
Regression Weights
Standardized Regression W
Variances
Squared Multiple Correlat
Matrices
Factor Score Weights
Total Effects
Standardized Total Effect
Direct Effects
Standardized Direct Effec
Indirect Effects
Standardized Indirect Effe
Covariances among Estimate
Correlations among Estimate
Critical Ratios for Difference
Fit
Fit Measures 1
Fit Measures 2
Execution Time

Estimates

Group number 1

Default model

Fit Measures

Fit Measure	Default model	Saturated	Independence	Macro
Discrepancy	6.755	0.000	203.080	CMIN
Degrees of freedom	5	0	10	DF
P	0.240		0.000	P
Number of parameters	10	15	5	NPAR
Discrepancy / df	1.351		20.308	CMINDF
RMR	0.089	0.000	1.574	RMR
GFI	0.957	1.000	0.349	GFI
Adjusted GFI	0.871		0.024	AGFI
Parsimony-adjusted GFI	0.319		0.233	PGFI
Normed fit index	0.967	1.000	0.000	NFI
Relative fit index	0.933		0.000	RFI
Incremental fit index	0.991	1.000	0.000	IFI
Tucker-Lewis index	0.982		0.000	TLI
Comparative fit index	0.991	1.000	0.000	CFI
Parsimony ratio	0.500	0.000	1.000	PRATIO
Parsimony-adjusted NFI	0.483	0.000	0.000	PNFI
Parsimony-adjusted CFI	0.495	0.000	0.000	PCFI
Noncentrality parameter estimate	1.755	0.000	193.080	NCP
NCP lower bound	0.000	0.000	150.439	NCPLO
NCP upper bound	12.801	0.000	243.156	NCPHI
FMIN	0.116	0.000	3.501	FMIN
F0	0.030	0.000	3.329	F0
F0 lower bound	0.000	0.000	2.594	F0LO
F0 upper bound	0.221	0.000	4.192	F0HI
RMSEA	0.078		0.577	RMSEA
RMSEA lower bound	0.000		0.509	RMSEALO
RMSEA upper bound	0.210		0.647	RMSEAHI
P for test of close fit	0.317		0.000	PCLOSE
Akaike information criterion (AIC)	26.755	30.000	213.080	AIC
Browne-Cudeck criterion	29.063	33.462	214.234	BCC
Bayes information criterion	63.625	85.305	231.515	BIC
Consistent AIC	57.530	76.163	228.468	CAIC
Expected cross validation index	0.461	0.517	3.674	ECVI
ECVI lower bound	0.431	0.517	2.939	ECVILO
ECVI upper bound	0.652	0.517	4.537	ECVIHI
MECVI	0.501	0.577	3.694	MECVI
Hoelter .05 index	96		6	HFIVE
Hoelter .01 index	130		7	HONE

Maximum Likelihood - a method for obtaining an estimate of unknown parameter of an assumed population distribution.

Output 6.4.C *Text Output*

Select the **View/Set...Text Output** menu path to view output as a single text document. In tables below C.R. signifies Critical Ratio, the estimate divided by its standard error. Critical Ratios greater than 1.96 are considered significant to the .05 level.

Maximum Likelihood Estimates

Regression Weights:	Estimate	S.E.	C.R.	Label
col2 <------ Status	1.766	0.204	8.659	par-1
col3 <------ Status	1.143	0.151	7.576	par-2
col4 <------ Status	1.235	0.201	6.131	par-3
col5 <------ Status	1.565	0.207	7.565	par-4
col1 <------ Status	1.191	0.152	7.827	par-5

Standardized Regression Weights:	Estimate
col2 <------ Status	0.901
col3 <------ Status	0.829
col4 <------ Status	0.718
col5 <------ Status	0.828
col1 <------ Status	0.846

Variances:	Estimate	S.E.	C.R.	Label
Status	1.000			
err_2	0.727	0.213	3.405	par-6
err_4	1.437	0.296	4.860	par-7
err-5	1.125	0.257	4.381	par-8
err_3	0.594	0.137	4.336	par-9
err_1	0.561	0.134	4.188	par-10

Squared Multiple Correlations:	Estimate
col1	0.717
col3	0.688
col5	0.685
col4	0.515
col2	0.811

Factor Score Weights

	col1	col3	col5	col4	col2
Status	0.160	0.145	0.105	0.065	0.183

Total Effects	Status
col1	1.191
col3	1.143
col5	1.565
col4	1.235
col2	1.766

Standardized Total Effects	Status
col1	0.846
col3	0.829
col5	0.828
col4	0.718
col2	0.901

7 Multidimensional Scaling

Sections 7.1 and 7.2 demonstrate the SPSS ALSCAL multidimensional scaling command. Section 7.3 contains MDS examples created by James Lattin using the KYST, SINDSCAL and MDPREF algorithms. These applications are available on the Analyzing Multivariate Data CD ROM.

Example7.1.A *Classical metric MDS of distances between European cities*

Define the working data file with the DATA LIST command, specifying column locations of distance terms. City distances values can be typed or copy-pasted between the BEGIN DATA – END DATA keywords. Read data values into the working data file by highlighting and running syntax beginning with DATA LIST through the END DATA keywords to.

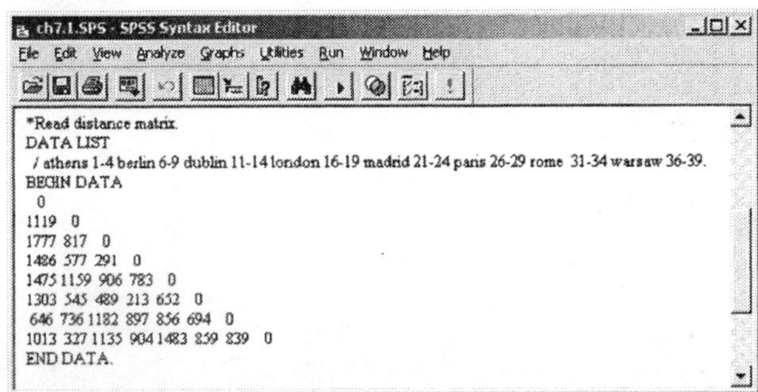


```
*Read distance matrix.
DATA LIST
 / athens 1-4 berlin 6-9 dublin 11-14 london 16-19 madrid 21-24 paris 26-29 rome 31-34 warsaw 36-39.
BEGIN DATA
 0
1119 0
1777 817 0
1486 577 291 0
1475 1159 906 783 0
1303 545 489 213 652 0
 646 736 1182 897 856 694 0
1013 327 1135 904 1483 859 839 0
END DATA.
```

Output 7.1.A: *Distances between European cities read into the working data file*

Output1 - SPSS Viewer

File Edit View Insert Format Analyze Graphs Utilities Window Help

Output
Log

```
*Read distance matrix.
DATA LIST
  / athens 1-4 berlin 6-9 dublin 11-14 london 16-19 madrid 21-24 paris 26-29 rom
  e 31-34 warsaw 36-39.

Data List will read 1 records from the command file

Variable   Rec   Start     End        Format

ATHENS      1       1       4          F4.0
BERLIN      1       6       9          F4.0
DUBLIN      1      11      14          F4.0
LONDON      1      16      19          F4.0
MADRID      1      21      24          F4.0
PARIS       1      26      29          F4.0
ROME        1      31      34          F4.0
WARSAW      1      36      39          F4.0

BEGIN DATA
   0
1119    0
1777  817    0
1486  577  291    0
1475 1159  906  783    0
1303  545  489  213  652    0
 646  736 1182  897  856  694    0
1013  327 1135  904 1483  859  839    0
END DATA.
```

Example7.1.B *Define Multidimensional Scaling Analysis*

Choose the path **Analyze-Scale-Multidimensional Scaling** to open the **Multidimensional Scaling** dialog box, include all 8city variables in the analysis.

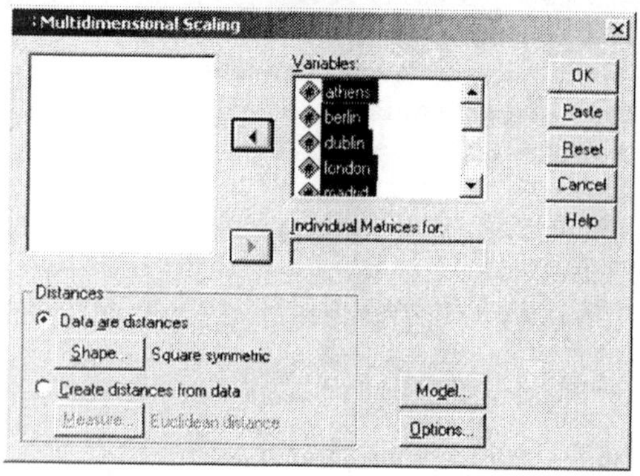

Open the **Model** dialog box and select interval level measurement. Other options remain default. Open the **Options** dialog box. Select **Group plots** and **Model and Options Summary**.

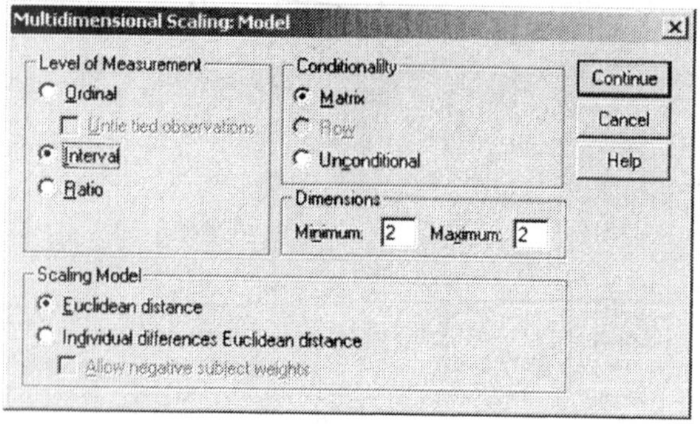

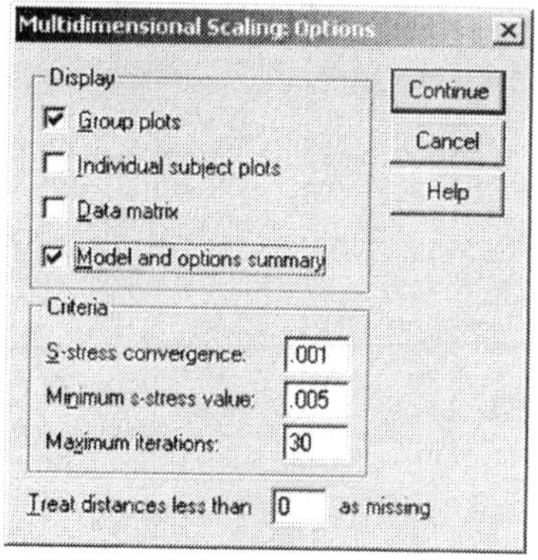

Pasted Syntax

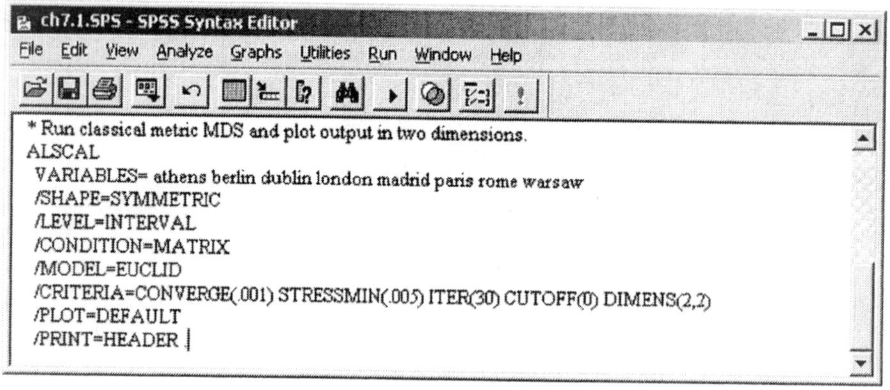

```
* Run classical metric MDS and plot output in two dimensions.
ALSCAL
  VARIABLES= athens berlin dublin london madrid paris rome warsaw
  /SHAPE=SYMMETRIC
  /LEVEL=INTERVAL
  /CONDITION=MATRIX
  /MODEL=EUCLID
  /CRITERIA=CONVERGE(.001) STRESSMIN(.005) ITER(30) CUTOFF(0) DIMENS(2,2)
  /PLOT=DEFAULT
  /PRINT=HEADER
```

Output 7.1.B: *Results from classical metric MDS of cities in Europe*

Alscal produces voluminous output, including details of analysis options, iteration history and plot coordinates

```
Alscal Procedure Options

Data Options-

Number of Rows (Observations/Matrix).    8
Number of Columns (Variables) . . .      8
Number of Matrices   . . . . . .         1
Measurement Level . . . . . . .          Interval
Data Matrix Shape . . . . . . .          Symmetric
Type  . . . . . . . . . . . . .          Dissimilarity
Approach to Ties  . . . . . . .          Leave Tied
Conditionality . . . . . . . .           Matrix
Data Cutoff at . . . . . . . .            .000000


Model Options-

Model . . . . . . . . . . . .            Euclid
Maximum Dimensionality  . . . . .        2
Minimum Dimensionality  . . . . .        2
Negative Weights  . . . . . . .          Not Permitted

Output Options-

Job Option Header . . . . . . .          Printed
Data Matrices  . . . . . . . .           Not Printed
Configurations and Transformations  .    Plotted
Output Dataset . . . . . . . .           Not Created
Initial Stimulus Coordinates  . . .      Computed

Algorithmic Options-

Maximum Iterations  . . . . . .            30
Convergence Criterion  . . . . .          .00100
Minimum S-stress  . . . . . . .           .00500
Missing Data Estimated by  . . . .       Ulbounds

Iteration history for the 2 dimensional solution (in squared distances)

                  Young's S-stress formula 1 is used.

                Iteration       S-stress        Improvement

                    1            .01490
                    2            .01254           .00236
                    3            .01230           .00024

                       Iterations stopped because
                 S-stress improvement is less than    .001000


           Stress and squared correlation (RSQ) in distances

RSQ values are the proportion of variance of the scaled data (disparities)
           in the partition (row, matrix, or entire data) which
            is accounted for by their corresponding distances.
             Stress values are Kruskal's stress formula 1.

                  For  matrix     Stress  =   .00958      RSQ =  .99951

           Configuration derived in 2 dimensions
```

Stimulus Coordinates

Stimulus Number	Stimulus Name	1	2
1	ATHENS	2.2017	-.5261
2	BERLIN	.1719	.8395
3	DUBLIN	-1.5570	.4007
4	LONDON	-.9454	.2500
5	MADRID	-.8828	-1.4908
6	PARIS	-.6033	-.0623
7	ROME	.8075	-.6418
8	WARSAW	.8074	1.2308

Output 7.1.C: *Scatter plot of metric MDS configuration for cities in Europe*

Derived Stimulus Configuration

Euclidean distance model

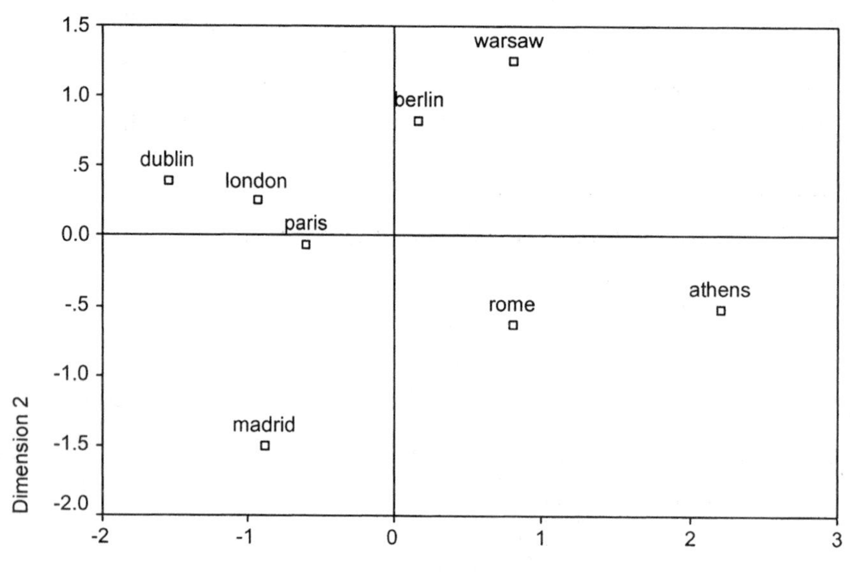

Example 7.2: *Nonmetric MDS of car dissimilarity data*

Define the working data file as in Example 7.1.A by running the DATA LIST, BEGIN DATA and the END DATA commands. Choose path **Analyze-Scale-Multidimensional Scaling** as in Example 7.1.B to define the analysis. Ordinal level analysis is the ALSCAL default, so the syntax is simpler than Example 7.1.

```
ch7.2.SPS - SPSS Syntax Editor
File Edit View Analyze Graphs Utilities Run Window Help

*Read distance matrix for Example 7.2.
DATA LIST
 / BMW 1-2 Ford_Exp 4-5 Infiniti 7-8 Jeep 10-11 Lexus 13-14 Chrysler 16-17 Mercedes 19-20 Saab 22-23 Porsche 25-26 Volvo 28-29.
BEGIN DATA
0
34 0
 8 24 0
31  2 25  0
 7 26  1 27  0
43 14 35 15 37  0
 3 28  5 29  4 42  0
10 18 20 17 13 36 19  0
 6 39 41 38 40 45 32 21  0
33 11 22 12 23  9 30 16 44  0
END DATA.

* 7.2: Run non- metric MDS and plot output in two dimensions.
ALSCAL
 VARIABLES= BMW Ford_Exp Infiniti Jeep Lexus Chrysler Mercedes Saab Porsche Volvo
/PLOT=DEFAULT
 /PRINT=HEADER .
```

Output 7.2.A: *Results from ALSCAL nonmetric MDS of car dissimilarity data*

```
Alscal Procedure Options

Data Options-

Number of Rows (Observations/Matrix).   10
Number of Columns (Variables) . . .     10
Number of Matrices    . . . . . . .      1
Measurement Level . . . . . . . . .      Ordinal
Data Matrix Shape . . . . . . . . .      Symmetric
Type  . . . . . . . . . . . . . . .      Dissimilarity
Approach to Ties    . . . . . . . .      Leave Tied
Conditionality . . . . . . . . . .       Matrix
Data Cutoff at . . . . . . . . . .        .000000

Model Options-

Model . . . . . . . . . . . . . . .      Euclid
Maximum Dimensionality  . . . . . .      2
Minimum Dimensionality  . . . . . .      2
Negative Weights  . . . . . . . . .      Not Permitted

Output Options-

Job Option Header . . . . . . . . .      Printed
Data Matrices   . . . . . . . . . .      Printed
Configurations and Transformations .     Plotted
Output Dataset . . . . . . . . . .       Not Created
Initial Stimulus Coordinates  . . .      Computed
```

```
Algorithmic Options-

Maximum Iterations  . . . . . . .      30
Convergence Criterion . . . . . .    .00100
Minimum S-stress  . . . . . . . .    .00500
Missing Data Estimated by . . . .   Ulbounds
Tiestore . . . . . . . . . . . .   45
Iteration history for the 2 dimensional solution (in squared distances)

                  Young's S-stress formula 1 is used.

                Iteration      S-stress       Improvement

                    1           .11829
                    2           .09764           .02065
                    3           .08800           .00964
                    4           .08025           .00775
                    5           .07346           .00679
                    6           .06761           .00585
                    7           .06274           .00487
                    8           .05898           .00376
                    9           .05630           .00268
                   10           .05461           .00169
                   11           .05352           .00109
                   12           .05272           .00080

                         Iterations stopped because
                 S-stress improvement is less than    .001000


            Stress and squared correlation (RSQ) in distances

RSQ values are the proportion of variance of the scaled data (disparities)
           in the partition (row, matrix, or entire data) which
            is accounted for by their corresponding distances.
             Stress values are Kruskal's stress formula 1.


                For  matrix    Stress  =   .04703      RSQ =  .98733


Configuration derived in 2 dimensions

                   Stimulus Coordinates

                        Dimension

Stimulus   Stimulus     1         2
 Number      Name

    1      BMW        1.3408   -.2430
    2      FORD_EXP  -1.0937   -.4542
    3      INFINITI    .4774    .8867
    4      JEEP      -1.0697   -.4565
    5      LEXUS       .6608    .8907
    6      CHRYSLER  -2.0814    .3915
    7      MERCEDES   1.2098    .5587
    8      SAAB        .2640   -.4411
    9      PORSCHE    1.3645  -1.7838
   10      VOLVO     -1.0727    .6510
```

Output 7.2.B: *Scatter plot of two-dimensional nonmetric MDS solution for car dissimilarity data and transformation scatterplot showing goodness-of-fit.*

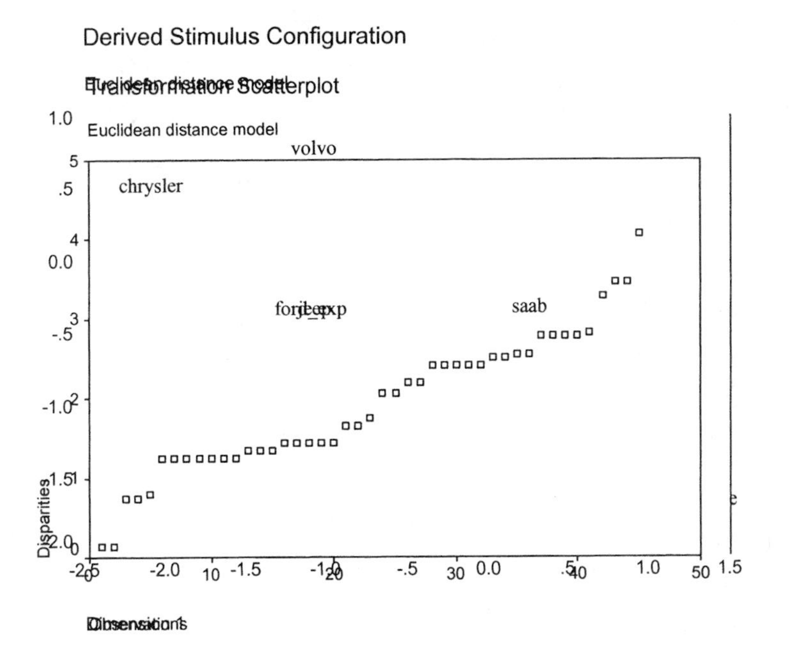

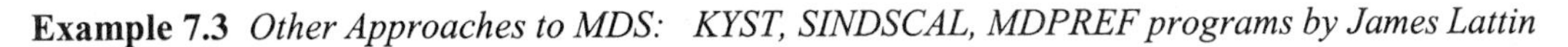

Example 7.3 *Other Approaches to MDS: KYST, SINDSCAL, MDPREF programs by James Lattin*

KYST Example 7.3.1 *Nonmetric MDS analysis of car dissimilarity data*

```
REGRESSION = ASCENDING  PRINT DATA RANDOM=88
ITERATIONS=200 STRMIN=0.0 CARDS SRATST=1.00
PRINT DISTANCES DIAGONAL ABSENT LOWERHALFMATRIX
DATA
DATA FOR 10 CARS. R=2
010001001
(9F3.0)
 34
  8 24
 31  2 25
  7 26  1 27
 43 14 35 15 37
  3 28  5 29  4 42
 10 18 20 17 13 36 19
  6 39 41 38 40 45 32 21
 33 11 22 12 23  9 30 16 44
COMPUTE
STOP
```

KYST Output 7.3.1.A *Convergence results*

```
HISTORY OF COMPUTATION. N=   10.     THERE ARE    45   DATA VALUES, SPLIT
INTO   1    LISTS.      DIMENSION =   2
ITERATION STRESS    SRAT SRATAV CAGRGL  COSAV  ACSAV    SFGR    STEP

        0   .497   .800   .800   .000   .000   .000   .0259   .3224
        1   .372   .748   .782   .788   .520   .520   .0094   .7099
        2   .369   .992   .847  -.411  -.094   .448   .0118   .5107
        3   .343   .928   .873  -.563  -.404   .524   .0088   .2664
        4   .317   .926   .890  -.394  -.398   .438   .0044   .1456
        5   .306   .966   .915   .079  -.083   .201   .0032   .1112
        6   .298   .972   .933   .403   .238   .335   .0033   .1302
<... data omitted ...>
       84   .078  1.000  1.000   .734   .702   .723   .0000   .0000
       85   .078  1.000  1.000  -.294   .044   .440   .0000   .0000
       86   .078  1.000  1.000  -.745  -.477   .642   .0000   .0000
MINIMUM WAS ACHIEVED
```

KYST Output 7.3.1.B *Results from nonmetric MDS of car dissimilarity data*

```
THE FINAL CONFIGURATION HAS BEEN ROTATED TO PRINCIPAL COMPONENTS.
THE FINAL CONFIGURATION OF 10 POINTS IN 2 DIMENSIONS HAS STRESS .078 FORMULA 1

 LABEL FOR CONFIGURATION PLOTS          FINAL CONFIGURATION
                                              1       2
                  A                     1  -.908    .060
                  B                     2   .809    .091
                  C                     3  -.152    .711
                  D                     4   .796    .073
                  E                     5  -.182    .667
                  F                     6  1.458   -.362
                  G                     7  -.695    .588
                  H                     8  -.252   -.348
                  I                     9 -1.451   -.915
                  J                    10   .577   -.565
```

KYST Output 7.3.2.C *Actual and fitted distances from nonmetric MDS of car dissimilarity data*

```
  I   J   DATA    DIST DHAT WGHT GP  NO
  5   3   1.000   .05  .04 1.00  1   1,
  4   2   2.000   .02  .04 1.00  1   2,
  7   1   3.000   .57  .54 1.00  1   3,
  7   5   4.000   .52  .54 1.00  1   4,
  7   3   5.000   .56  .56 1.00  1   5,
  9   1   6.000  1.12  .87 1.00  1   6,
  5   1   7.000   .95  .87 1.00  1   7,
  3   1   8.000  1.00  .87 1.00  1   8,
 10   6   9.000   .90  .87 1.00  1   9,
  8   1  10.000   .77  .87 1.00  1  10,

<... data omitted ...>

  9   3  41.000  2.08 2.26 1.00  1  41,
  7   6  42.000  2.35 2.27 1.00  1  42,
  6   1  43.000  2.40 2.27 1.00  1  43,
 10   9  44.000  2.06 2.27 1.00  1  44,
  9   6  45.000  2.96 2.96 1.00  1  45,
```

SINDSCAL Example 7.3.2 *Individual differences scaling model of breakfast food dissimilarity data from four subjects*

```
  2   2   4  15
100   7   0   0  76
Breakfast Data 4 Subjects 15 Stim
(4X,15F4.0)
  42   0  15  25   3  14  24  28   7   8  16  26  21  20  16  27
  42  15   0  15  24   3  17   8   7   6   2  17  25  18  22  11
  42  25  15   0  22  17   2   4  20  21  16  10  11  24  11   3
  42   3  24  22   0  22  21  27  11  12  25  17   5   2  13  26

<… remaining data for subject 42 omitted …>

  71   0  20  22  11  10   5  22  10  11  18  14   7   8   9  12
  71  20   0   5  24  19  20   6  18  14   3  23  25  27  24  19
  71  22   5   0  26  20  21   4  17  16   3  22  24  25  23  16
  71  11  24  26   0  16  17  28  16  19  27   6   4   2   5  19

<… remaining data for subject 71 omitted …>

 101   0  21  15   8  16  15  24  11  10  24   7   3   7   9  13
 101  21   0   6  27  20  12   3   7   6   3  24  23  24  22  19
 101  15   6   0  28  21   8   4  12  11   2  25  25  26  24  21
 101   8  27  28   0  11  21  25  22  24  26   4   8   2   8  13

<… remaining data for subject 101 omitted …>

 112   0  15  16  17   9  24  20   8  10  13  21  10  17  22  20
 112  15   0   7  21   5  20  11   7   6   2  27  26  24  23  11
 112  16   7   0  25  12   8   9  12  11   5  24  21  21  19   2
 112  17  21  25   0  25  17  27  15  13  24   6   4   2   8  27

<… remaining data for subject 112 omitted …>
```

SINDSCAL Output 7.3.2.A *Program parameters and initial values*

```
                       SYMMETRIC INDSCAL
 Breakfast Data 4 Subjects 15 Stim
 **************************************************
  PARAMETERS
   DIM  IRDATA  ITMAX IPUNCH  IPLOT    IRN
    2      7     100     0      0       76
  NO. OF MATRICES =   4  NO. OF STIM. =  15
 **************************************************
 INITIAL STIMULUS MATRIX
    1   .339   .023   .215  -.215   .005   .431   .272   .272   .078   .120
       -.015   .289   .056  -.409  -.146
 2   .338   .441   .343   .380  -.082  -.268  -.124   .061  -.471   .429
       -.309   .399  -.328   .033   .286
```

SINDSCAL Output 7.3.2.B *Convergence results for SINDSCAL*

```
                          HISTORY OF COMPUTATION
  ITERATION          CORRELATIONS BETWEEN        VAF              LOSS
                      Y(DATA) AND YHAT         (R**2)          (Y-YHAT)**2
       0                  .047442              .002251           .997749
       1                  .465083              .216302           .813527
       2                  .785665              .617269           .384639
```

```
       3              .844654            .713441            .286638
       4              .866503            .750828            .249250
       5              .877035            .769191            .230825
       6              .878767            .772231            .227770
       7              .879113            .772840            .227160
       8              .879265            .773108            .226892
       9              .879283            .773138            .226862
      10              .879292            .773155            .226845
      11              .879294            .773157            .226843
      12              .879294            .773158            .226842
REACHED CRITERION ON ITERATION 12
    FINAL             .879294            .773158            .226842
```

SINDSCAL Output 7.3.2.C *Results from individual differences scaling model of breakfast food dissimilarity data*

```
                         NORMALIZED SOLUTION
SUBJECTS WEIGHT MATRIX
  1      .560   .889   .855   .741
  2      .676   .087   .135   .472
STIMULUS MATRIX
  1     -.129   .349   .323  -.368   .006   .029   .348   .091   .118   .337
        -.273  -.314  -.332  -.271   .087
  2      .378   .166  -.243   .153   .237  -.338  -.254   .355   .322   .164
        -.229  -.137   .020  -.238  -.357
NORMALIZED SUM OF PRODUCTS (SUBJECTS)
  1     1.000
  2      .710  1.000
SUM OF PRODUCTS (STIMULI)
  1     1.000
  2      .035  1.000
APPROXIMATE PROPORTION OF TOTAL VARIANCE ACCOUNTED FOR BY EACH DIMENSION
          1      2
        .597   .177
CORRELATION BETWEEN COMPUTED SCORES AND SCALAR PROD. FOR SUBJECTS
   1      .878373
   2      .893290
   3      .865732
   4      .879565
```

MDPREF Example7.3.3 *Multidimensional preference model of movie critic data*

```
   6  18   6   3   2   0
 (18F4.0)
   6   7  10  11   9  10   9   5   3  10   7   9  12   8   9   3   6   9
  12   8  11  13  10  11   9   9  10   8   6   9  10   5   7   9   7   3
   8   9   8   8   9   9   8   7  10   7   7   9   9   7  10   8   9  10
   7  10   8  12   6   7   6   3   5   9   9   8   5   6   8   7   9  11
   6  10  11   6  10   7   6   5   9   7   5  10  11   5   6   6   6   6
   6   9   7   9   7   9   7   8  11   9   8  10   6  10  10   8   5   8
Ebert
Bernard
Rickey
Clark
Kempley
Turan
AR
AP2
DK
EWS

<… remaining movie labels omitted …>
```

MDPREF Output 7.3.3.A *Parameter values for MDPREF*

```
                              M D P R E F
                 MULTIDIMENSIONAL ANALYSIS OF PREFERENCE DATA
             PROGRAM WRITTEN BY DR. J. D. CARROLL AND JIH JIE CHANG
                  Text Version for Analyzing Multivariate Data
            by James M. Lattin, J. Douglas Carroll and Paul E. Green
                         PC - MDS VERSION,  2003


ANALYSIS START:  DATE  10/10/2002,  TIME  16:15:13

ANALYSIS TITLE: Movies and Critics
DATA IS READ FROM FILE: c:pref_in.txt
Output FILE IS: c:pref_out.txt

NP (NO. OF DATA ROWS (VECTORS OR SUBJECTS))                     6
NS (NO. OF COLUMNS (POINTS OR STIMULI))                        18
NF (NO. OF DIMENSIONS)                                          6
NFP (NO. OF DIMENSIONS PLOTTED)                                 3

IREAD  1=NP X NS SCORE MATRIX WITH ROW MEAN SUBTRACTED          2
       2=SAME AS 1 WITH SCORES DIVIDED BY ROW S. D.

NORP   0=NORMALIZE SUBJ. VECTORS                                0
       1=DO NOT
```

MDPREF Output 7.3.3.B

```
                      ROOTS OF THE FIRST SCORE MATRIX
  29.6531      24.8088      21.2649      14.9455      10.9775      6.3501

                    PROPORTION OF VARIANCE ACCOUNTED FOR BY EACH FACTOR
     1            2            3            4            5            6
  .2746        .2297        .1969        .1384        .1016        .0588

                    CUMULATIVE PROPORTION OF VARIANCE ACCOUNTED FOR
     1            2            3            4            5            6
  .2746        .5043        .7012        .8396        .9412       1.0000
```

MDPREF Output 7.3.3C

POPULATION MATRIX (ROWS OR VECTORS)

FACTOR						
Ebert	.5900	.2872	-.5576	.2338	-.3860	-.2342
Bernard	.5140	-.6089	-.0733	.3478	.4631	-.1557
Rickey	.5114	.4087	.5535	-.4071	.1308	-.2867
Clark	.1232	.7634	-.4077	.0379	.4471	.1856
Kempley	.8614	-.1258	.2413	-.0277	-.1619	.3962
Turan	-.1281	.3993	.5782	.6976	-.0559	-.0056

MDPREF Output 7.3.3.D

NORMALIZED STIMULUS MATRIX (COLUMNS OR POINTS)

FACTOR						
AR	-.0559	-.3236	-.1499	-.1383	.3624	-.1278
AP2	.1520	.2053	.1386	-.0111	.1333	.4566
DK	.3384	-.1495	-.1727	.0401	-.0351	.3450
EWS	.1497	.1604	-.3455	.4386	.4889	-.2100
AIH	.2718	-.1662	.0386	-.1477	-.1270	.0230

<… scores for remaining movies omitted …>

MDPREF Output 7.3.3.E

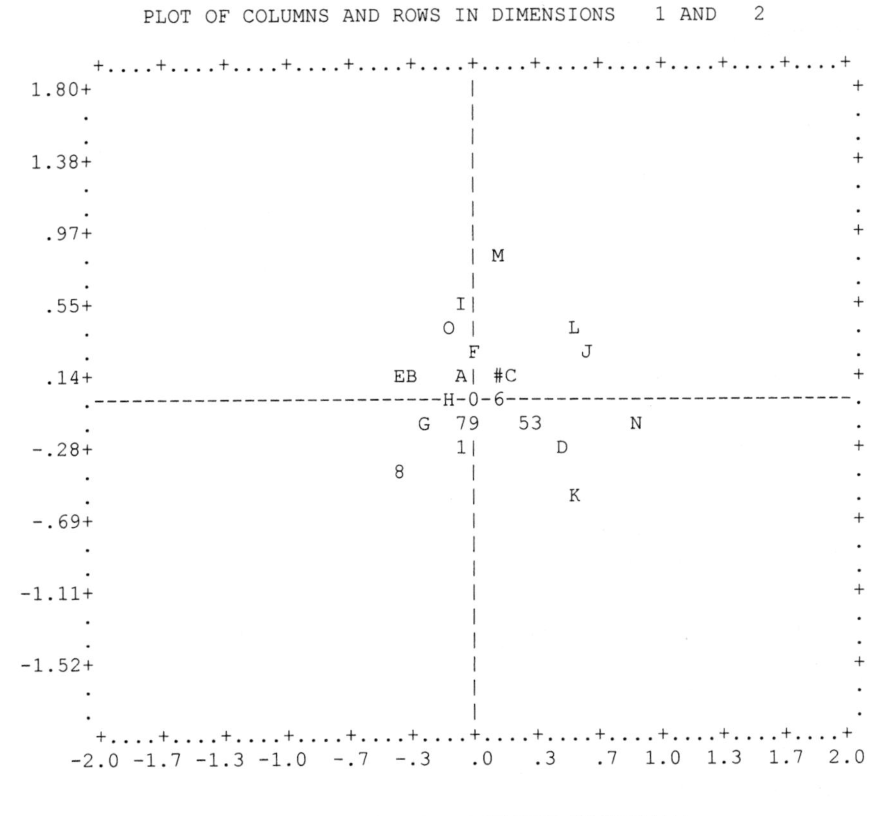

*****IDENTIFICATION KEY FOR PLOTS WITH IDENTIFIED POINTS*****

PT #	1	2	3	4	5	6	7	8	9	10	11	12	13	14	15
CHAR	1	2	3	4	5	6	7	8	9	A	B	C	D	E	F

PT #	16	17	18	19	20	21	22	23	24	25	26	27	28	29	30
CHAR	G	H	I	J	K	L	M	N	O	P	Q	R	S	T	U

IN JOINT SPACE PLOTS, THE FIRST 18 POINTS ARE COLUMNS AND THE NEXT 6 ARE ROW (VECTOR) END POINTS.

8 Cluster Analysis Using SPSS

8.1 Single-linkage cluster of bimodal data
8.2 Ward's method and K-means analysis of preference segmentation
8.3 Validation of cluster analysis

The examples in this chapter demonstrate hierarchical cluster analysis, K-means clustering, creating dendrograms to visualize relationships, saving cluster membership as variables for further analysis, and validation of cluster analysis by cross-tabulation of parallel samples.

Example 8.1: *Single-linkage cluster analysis of bimodal sample data*

This example corresponds to the textbook discussion on agglomerative clustering in Section 8.4. Open Bimodal.sav as the active data file. Click on the menu path **Analyze-Classify** and choose **Hierarchical Cluster Analysis**. Select col1 and col2 as variables for analysis. Click on the **Plots** button and select **Dendrogram**.

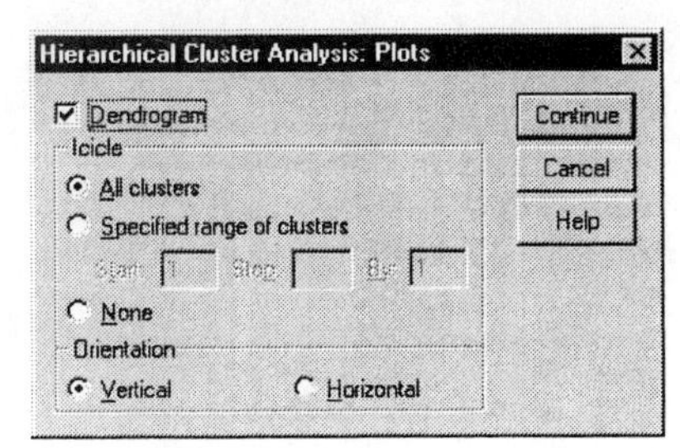


```
*Open file bimodal.sav by double-clicking icon.
*8.1 Single-linkage cluster analysis of bimodal sample data.
CLUSTER col1 col2
 /METHOD single
 /PRINT SCHEDULE
 /PLOT DENDROGRAM .
```

Output 8.1 *Dendrogram using Single Linkage*

```
Num  +---------+---------+---------+---------+---------+

 26   ⇩×⇩⇩⇩⇘
 28   ⇩⇗    ⇔
 27   ⇩⇩⇩⇘ ⇔
 42   ⇩⇩⇩⇗ ⇔
 32   ⇩⇩⇩✓⇩×⇩⇩⇩⇘
 39   ⇩⇩⇩⇗ ⇔    ⇔
 48   ⇩⇩⇩⇩⇩⇗    ▫⇩⇩⇩⇘
 45   ⇩⇩⇩×⇩⇗    ⇔    ⇔
 50   ⇩⇩⇩⇗      ⇔    ⇔
 49   ⇩⇩⇩⇩⇩⇩⇩⇩⇩⇩⇗   ▫⇩⇩⇩⇘
 30   ⇩⇩⇩×⇩⇩⇩⇩⇩⇘   ⇔    ⇔
 34   ⇩⇩⇩⇗      ▫⇩⇘ ⇔    ⇔
 29   ⇩⇩⇩⇩⇩⇩⇩⇩⇩⇩⇗ ⇔ ⇔    ⇔
 31   ⇩⇩⇩⇩⇩⇘      ▫⇩⇗    ⇔
 33   ⇩⇩⇩⇩⇩✓⇩⇩⇩⇩⇩⇘       ⇔
 43   ⇩⇩⇩⇩⇩⇗       ⇔       ▫⇩⇘
 47   ⇩⇩⇩⇩⇩⇩⇩⇩⇩⇩⇩⇩⇗       ⇔ ⇔
 44   ⇩⇩⇩⇩⇩⇩⇩⇩⇩⇩⇩⇩⇩⇩⇩⇩⇩⇩⇩⇗ ▫⇩⇘
 35   ⇩⇩⇩⇩⇩⇩⇩⇩⇩⇩⇩⇩⇩⇩⇩⇩⇩⇩⇩⇗ ⇔ ⇔
 12   ⇩×⇩⇩⇩⇩⇩⇩⇩⇩⇘       ⇔ ⇔
 38   ⇩⇗       ▫⇩⇩⇩⇩⇩⇩⇩⇩⇩⇩⇗ ▫⇩⇩⇩⇩⇩⇩⇩⇩⇩⇩⇩⇩⇘
 41   ⇩⇩⇩⇩⇩⇩⇩⇩⇩⇩⇗        ⇔           ⇔
  1   ⇩⇩⇩×⇩⇩⇩⇩⇩⇩⇩⇩⇩⇩⇩⇩⇩⇘  ⇔           ▫⇩⇩⇩⇩⇩⇘
  6   ⇩⇩⇩⇗           ▫⇩⇩⇩⇗          ⇔      ⇔
 46   ⇩⇩⇩⇩⇩⇩⇩⇩⇩⇩⇩⇩⇩⇩⇩⇩⇩⇩⇩⇗          ⇔      ⇔
 37   ⇩⇩⇩⇩⇩⇩⇩⇩⇩⇩⇩⇩⇩⇩⇩⇩⇩⇩⇩⇩⇩⇩⇩⇩⇩⇩⇩⇩⇩⇩⇩⇩⇩⇩⇩⇩⇩⇗       ⇔
  9   ⇩⇩⇩×⇩⇘                         ⇔
 18   ⇩⇩⇩⇗ ▫⇩⇩⇩⇘                     ⇔
 21   ⇩⇩⇩⇩⇩⇗    ▫⇩⇩⇩⇩⇩⇩⇩⇩⇩⇩⇩⇩⇩⇩⇩⇩⇩⇘          ⇔
 10   ⇩⇩⇩⇩⇩⇩⇩⇩⇩⇩⇗          ⇔           ▫⇩⇩⇩⇩⇩⇩⇩⇩⇩⇩⇘
 15   ⇩⇩⇩⇩⇩×⇩⇘          ▫⇩⇩⇩⇩⇩⇩⇩⇩⇩⇩⇘  ⇔       ⇔
 20   ⇩⇩⇩⇩⇩⇗ ▫⇩⇘        ⇔          ⇔  ⇔       ⇔
  3   ⇩⇩⇩⇩⇩⇩⇩⇩⇗ ⇔        ⇔          ⇔  ⇔       ⇔
  8   ⇩⇩⇩⇩⇩⇩⇩×⇩×⇩⇩⇩⇩⇩⇩⇩⇩⇩⇩⇩⇩⇩⇩⇩⇩⇗      ⇔   ⇔       ⇔
 23   ⇩⇩⇩⇩⇩⇩⇩⇩⇗ ⇔                ⇔  ⇔       ⇔
 22   ⇩⇩⇩⇘      ⇔                ⇔  ⇔      ⇔
 24   ⇩⇩⇩✓⇩⇘    ⇔                ⇔  ⇔      ⇔
  7   ⇩⇩⇩⇗ ⇔    ⇔                ▫⇩⇩⇩⇗      ⇔
 16   ⇩⇩⇩⇩⇩✓⇩⇘ ⇔                 ⇔          ⇔
 14   ⇩⇩⇩⇩⇩⇗ ⇔ ⇔                 ⇔          ⇔
 11   ⇩×⇩⇘   ▫⇩⇗                 ⇔          ⇔
 25   ⇩⇗ ⇔    ⇔                  ⇔          ⇔
  4   ⇩⇩⇩✓⇩⇩⇩⇘                   ⇔          ⇔
  2   ⇩⇩⇩⇗    ⇔                  ⇔          ⇔
  5   ⇩⇩⇩⇩⇩⇩⇩⇗                   ⇔          ⇔
 17   ⇩⇩⇩⇩⇩⇩⇩⇩⇩⇩⇩⇩⇩⇩⇩⇩⇩⇩⇩⇩⇩⇩⇩⇩⇩⇩⇩⇩⇩⇩⇩⇩⇩⇩⇩⇩⇩⇩⇩⇗       ⇔
 13   ⇩⇩⇩⇩⇩⇩⇩⇩⇩⇩⇩⇩⇩⇩⇩⇩⇩⇩⇩⇩⇩⇩⇩⇩⇩⇩⇩×⇩⇩⇩⇩⇩⇩⇩⇩⇩⇩⇩⇩⇩⇩⇩⇩⇩⇩⇩⇩⇩⇩⇩⇩⇩⇩⇩⇗
 19   ⇩⇩⇩⇩⇩⇩⇩⇩⇩⇩⇩⇩⇩⇩⇩⇩⇩⇩⇩⇩⇩⇩⇩⇩⇩⇩⇩⇗                 ⇔
 36   ⇩⇩⇩⇩⇩⇩⇩⇩⇩⇩⇩⇩⇩⇩⇩⇩⇩⇩⇩⇩⇩⇩⇩⇩⇩⇩⇩×⇩⇩⇩⇩⇩⇩⇩⇩⇩⇩⇩⇩⇩⇩⇩⇩⇩⇩⇩⇩⇩⇩⇩⇩⇩⇩⇩⇗
 40   ⇩⇩⇩⇩⇩⇩⇩⇩⇩⇩⇩⇩⇩⇩⇩⇩⇩⇩⇩⇩⇩⇩⇩⇩⇩⇩⇩⇗
```

Example 8.2 *MBA car data: Ward's method and K-means*

This example corresponds to the textbook discussion on preference segmentation in section. Open mba_car_pref1.sav as the active data file. Example 8.2 has three parts: with SPSS CLUSTER command save centroids; with the MEANS command calculate cluster centers; run K-Means analysis with the QUICK CLUSTER command.

Ch 8 syntax.SPS - SPSS Syntax Editor

File Edit View Analyze Graphs Utilities Run Window Help

```
*8.2.
*mba_car_pref1.sav is the working data file (calibration sample n=152).
*Use Ward's method to identify and save five cluster centroids.
CLUSTER col2 col3 col4 col5 col6 col7 col8 col9 col10 col11
/METHOD ward
/SAVE=CLUSTERS (5)
/PLOT dendrogram
/PRINT schedule.

*Calculate initial cluster centers for K-Means clustering.
MEANS
 TABLES=col2 col3 col4 col5 col6 col7 col8 col9 col10 col11 BY clu5_1
 /CELLS mean .

*K-means analysis syntax.
*Copy-paste initial cluster centers from MEANS output into the /INITIAL subcommand.
QUICK CLUSTER col2 col3 col4 col5 col6 col7 col8 col9 col10 col11
/METHOD KMEANS(update)
/CRITERIA= clusters (5) MXITER(25) CONVERGE(.02)
 /INITIAL= (4.25 5.46    3.83    6.79    4.96    2.37    5.37    4.04    3.67    6.21
7.08    6.60    4.20    6.76    5.52    2.48    6.28    6.64    6.92    5.32
7.25    4.46    4.18    5.11    5.82    1.36    5.96    4.18    2.89    2.82
5.96    4.59    2.98    5.65    4.00    1.39    5.24    3.98    7.28    2.37
7.25    3.18    4.93    3.82    5.54    1.43    6.75    6.07    6.75    2.86)
/PRINT initial cluster id (col1) distance anova
/SAVE=cluster distance(dis).
*Note that col1 is the id variable.
```

Example 8.2.A *A five cluster hierarchical cluster analysis*

Open the menu path **Analyze-Classify-Hierarchical Cluster Analysis** as in Example 8.1. Select variables col2 through col11 for analysis. Open the **Hierarchical Cluster Analysis** dialog box. Specify Ward's method of analysis. Open the **Statistics** dialog box and under **Cluster Membership** specify a single solution of five clusters. A variable identifying cluster membership of each case will be saved automatically.

Hierarchical Cluster Analysis: Statistics

- [x] Agglomeration schedule
- [ ] Proximity matrix

Cluster Membership
- ( ) None
- (•) Single solution: 5 clusters
- ( ) Range of solutions:

From [ ] through [ ] clusters

Continue | Cancel | Help

Output 8.2.A: *Agglomeration Schedule from Ward's method. (Stages 13 through 139 not shown)*

The coefficients column measures distance between cases or clusters as they are joined step by step. The scale of coefficients varies according to the linkage method employed.

Agglomeration Schedule

Stage	Cluster Combined		Coefficients	Stage Cluster First Appears		Next Stage
	Cluster 1	Cluster 2		Cluster 1	Cluster 2	
1	82	95	1.5	0	0	39
2	41	53	3.5	0	0	11
3	113	119	6	0	0	80
4	49	116	8.5	0	0	78
5	71	115	12	0	0	18
6	137	152	16	0	0	35
7	79	104	20	0	0	37
8	101	102	24	0	0	30
9	20	123	28.5	0	0	33
10	22	30	33	0	0	52
11	41	124	37.667	2	0	56
12	54	133	42.667	0	0	62
....	....	....	....	....	....	....
140	3	4	3258.061	127	129	144
141	13	71	3398.02	137	108	147
142	2	29	3543.883	122	132	148
143	1	10	3691.214	139	113	146
144	3	16	3871.803	140	115	147
145	5	6	4066.241	136	138	149
146	1	7	4278.316	143	133	148
147	3	13	4600.798	144	141	149
148	1	2	4952.446	146	142	150
149	3	5	5402.296	147	145	150
150	1	3	6049.271	148	149	0

Example 8.2.B *Define the initial cluster seeds*

The objective of this step is to define the initial cluster seeds to be submitted to K-means cluster analysis. Use the MEANS command to save a table with mean values for all analysis variables, for each of the five clusters. Open the menu path **Analyze-Compare Means-Means**. Choose col2 through col11 as dependent variables and clu5_1. Click on the **Options** button to specify that only mean values are included in the output table (the N and std deviation are default included).

Output 8.2.B *SPSS MEANS defines the initial cluster seeds to be submitted to K-means cluster analysis*

Report

Mean

Ward Method	COL2	COL3	COL4	COL5	COL6	COL7	COL8	COL9	COL10	COL11
1	4.25	5.46	3.83	6.79	4.96	2.37	5.37	4.04	3.67	6.21
2	7.08	6.60	4.20	6.76	5.52	2.48	6.28	6.64	6.92	5.32
3	7.25	4.46	4.18	5.11	5.82	1.36	5.96	4.18	2.89	2.82
4	5.96	4.59	2.98	5.65	4.00	1.39	5.24	3.98	7.28	2.37
5	7.25	3.18	4.93	3.82	5.54	1.43	6.75	6.07	6.75	2.86
Total	6.35	4.77	3.90	5.58	5.03	1.73	5.85	4.85	5.74	3.64

Example 8.2.C *K-means cluster analysis*
Open the **Analyze-Classify-K-means Cluster Analysis** path. The SPSS command for K-means clustering is QUICK CLUSTER. Click the **Save** button to save cluster membership and distance from cluster center.

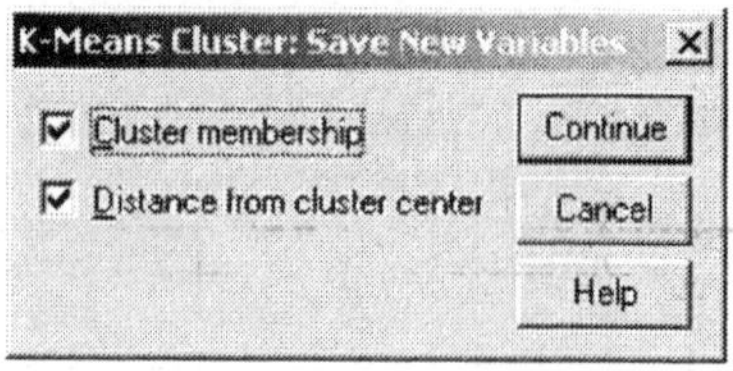

K-Means Cluster Analysis
col1
clu5_1
qcl_1
dis
Variables:
col2
col3
col4
col5
Label Cases by:
Number of Clusters: 5
Method
Iterate and classify
Classify only
OK
Paste
Reset
Cancel
Help
Centers >>
Iterate...
Save...
Options...

K-Means Cluster: Save New Variables
Cluster membership
Distance from cluster center
Continue
Cancel
Help

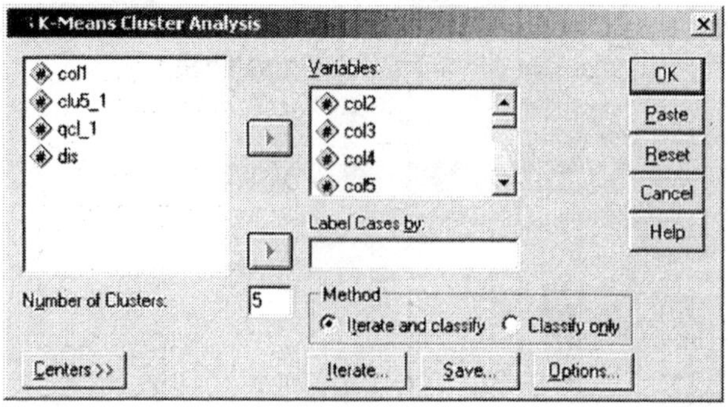

Paste the command into a syntax file. Within the syntax file directly type the /INITIAL = () subcommand following the /CRITERIA subcommand. Copy-paste initial cluster center values from Example 8.2.B MEANS output.

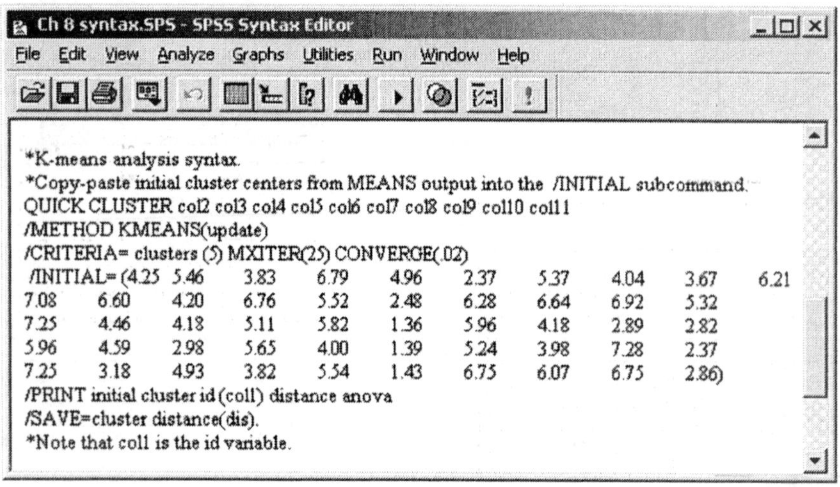

Ch 8 syntax.SPS - SPSS Syntax Editor
File Edit View Analyze Graphs Utilities Run Window Help

```
*K-means analysis syntax.
*Copy-paste initial cluster centers from MEANS output into the /INITIAL subcommand.
QUICK CLUSTER col2 col3 col4 col5 col6 col7 col8 col9 col10 col11
/METHOD KMEANS(update)
/CRITERIA= clusters (5) MXITER(25) CONVERGE(.02)
/INITIAL= (4.25  5.46  3.83  6.79  4.96  2.37  5.37  4.04  3.67  6.21
7.08  6.60  4.20  6.76  5.52  2.48  6.28  6.64  6.92  5.32
7.25  4.46  4.18  5.11  5.82  1.36  5.96  4.18  2.89  2.82
5.96  4.59  2.98  5.65  4.00  1.39  5.24  3.98  7.28  2.37
7.25  3.18  4.93  3.82  5.54  1.43  6.75  6.07  6.75  2.86)
/PRINT initial cluster id (col1) distance anova
/SAVE=cluster distance(dis).
*Note that col1 is the id variable.
```

Output 8.2.C *K-means cluster results*
Each cell in the Iteration History table indicates the distance that a cluster center has moved from its location in the previous iteration. The Distance column in the Cluster Membership table indicates how far each case is from its cluster center. (Cases 18 through 140 omitted)

Initial Cluster

	Cluste				
	1	2	3	4	5
COL	4.2	7.0	7.2	5.9	7.2
COL	5.4	6.6	4.4	4.5	3.1
COL	3.8	4.2	4.1	2.9	4.9
COL	6.7	6.7	5.1	5.6	3.8
COL	4.9	5.5	5.8	4.0	5.5
COL	2.3	2.4	1.3	1.3	1.4
COL	5.3	6.2	5.9	5.2	6.7
COL	4.0	6.6	4.1	3.9	6.0
COL1	3.6	6.9	2.8	7.2	6.7
COL1	6.2	5.3	2.8	2.3	2.8

Input from INITIAL

Iteration History [a]

	Change in Cluster Centers				
Iteration	1	2	3	4	5
1	1.540	1.174	.585	1.600	1.546
2	.944	1.020	.499	2.547	1.326
3	7.260E-02	.295	.655	.134	.630
4	5.584E-03	7.575E-03	1.598E-02	7.056E-03	1.432E-02

a. Convergence achieved due to no or small distance change. The maximum distance by which any center has changed is 8.064E-03. The current iteration is 4. The minimum distance between initial centers is 4.511.

Final Cluster Centers

	Cluster				
	1	2	3	4	5
COL2	4	6	7	4	7
COL3	5	6	4	5	4
COL4	4	4	5	3	4
COL5	7	7	5	5	5
COL6	5	5	6	3	5
COL7	3	2	2	2	1
COL8	6	6	6	4	7
COL9	4	6	4	3	5
COL10	2	7	3	5	8
COL11	7	5	3	2	2

Distances between Final Cluster Centers

Cluster	1	2	3	4	5
1		5.931	6.185	7.027	8.490
2	5.931		5.312	6.314	4.671
3	6.185	5.312		5.753	4.798
4	7.027	6.314	5.753		5.442
5	8.490	4.671	4.798	5.442	

Cluster Membership

Case Number	Cluster	Distance
1	4	4.786
2	5	6.829
3	5	4.570
4	3	5.718
5	5	3.718
6	2	4.583
7	4	3.882
8	3	7.765
9	.	.
10	2	6.044
11	2	5.768
12	5	3.545
13	5	6.376
14	5	5.534
15	5	5.639
16	3	3.693
17	1	5.109
....	....	....
141	4	6.528
142	4	5.052
143	4	4.337
144	5	4.443
145	2	4.449
146	4	3.869
147	4	4.072
148	1	5.084
149	4	5.734
150	3	5.788
151	2	5.818

The Cluster columns in the ANOVA table display variance attributable to each cluster. F ratio size indicates a variable's importance in the pattern of cluster separation.

ANOVA

	Cluster		Error			
	Mean Square	df	Mean Square	df	F	Sig.
COL2	45.852	4	2.322	146	19.748	.000
COL3	29.059	4	4.083	146	7.117	.000
COL4	8.371	4	3.082	146	2.716	.032
COL5	24.615	4	3.660	146	6.725	.000
COL6	18.538	4	3.286	146	5.642	.000
COL7	8.054	4	1.655	146	4.866	.001
COL8	28.988	4	2.600	146	11.151	.000
COL9	35.559	4	3.798	146	9.362	.000
COL10	152.825	4	2.151	146	71.035	.000
COL11	99.846	4	2.420	146	41.260	.000

The F tests should be used only for descriptive purposes because the clusters have been chosen to maximize the differences among cases in different clusters. The observed significance levels are not corrected for this and thus cannot be interpreted as tests of the hypothesis that the cluster means are equal.

Number of Cases in each Cluster

Cluster	1	12.000
	2	39.000
	3	41.000
	4	17.000
	5	42.000
Valid		151.000
Missing		1.000

Example 8.3: *Validation of cluster analysis of MBA car data*

The objective of this example is to compare K-means clustering results using cluster centers based on the calibration and validation samples. The working data set is mba_car_pref2.sav, the validation sample.

```
8.3.SPS - SPSS Syntax Editor
File Edit View Analyze Graphs Utilities Run Window Help

*8.3 Validation of cluster analysis of MBA car data.

*Step 1, K-means analysis of the validation sample with initial cluster centers from the calibration sample.
*mba_car_pref2.sav is the working data file (n=151).
QUICK CLUSTER col2 col3 col4 col5 col6 col7 col8 col9 col10 col11
/METHOD KMEANS(update)
/CRITERIA= clusters (5) MXITER(25) CONVERGE(.02)
/INITIAL= (4.25 5.46   3.83   6.79   4.96   2.37   5.37   4.04   3.67   6.21
7.08   6.60   4.20   6.76   5.52   2.48   6.28   6.64   6.92   5.32
7.25   4.46   4.18   5.11   5.82   1.36   5.96   4.18   2.89   2.82
5.96   4.59   2.98   5.65   4.00   1.39   5.24   3.98   7.28   2.37
7.25   3.18   4.93   3.82   5.54   1.43   6.75   6.07   6.75   2.86)
/PRINT initial cluster id (col1) distance anova
/SAVE=cluster.

*Step 2, With the validation sample use Wards method to identify cluster centers.
*This replicates procedures previously run in Example 8.2 on calibration data.
CLUSTER col2 col3 col4 col5 col6 col7 col8 col9 col10 col11
/METHOD ward
/SAVE=CLUSTERS (5)
/PRINT CLUSTER (3,5) SCHEDULE.
*Calculate K-means initial centers.
MEANS
 TABLES=col2 col3 col4 col5 col6 col7 col8 col9 col10 col11 BY clu5_1
 /CELLS mean .
*K-means analysis .
*Copy-paste initial cluster centers from MEANS output into the /INITIAL subcommand.
QUICK CLUSTER col2 col3 col4 col5 col6 col7 col8 col9 col10 col11
/METHOD KMEANS(update)
/CRITERIA= clusters (5) MXITER(25) CONVERGE(.02)
/INITIAL= (6.64 6.32   2.68   6.84   4.12   1.52   5.76   2.64   6.84   2.04
7.16   3.81   5.21   4.44   6.49   1.91   7.16   3.81   6.37   3.42
7.33   4.07   2.93   4.33   3.44   1.30   6.04   6.52   6.41   2.11
7.26   5.41   4.85   5.68   6.26   3.09   6.82   6.85   7.03   5.26
4.50   6.33   4.33   6.06   4.61   3.22   4.00   5.94   2.72   5.11)
/PRINT initial cluster id (col1) distance anova
/SAVE=cluster.

*Step 3, Crosstabulate results of Steps 1 and 2.
CROSSTABS
 /TABLES=qcl_1 BY qcl_2
 /FORMAT= AVALUE TABLES
 /CELLS= COUNT ROW COLUMN TOTAL .
```

The first and second steps of Example 8.3 essentially repeat the syntax of Example 8.2 for calculating K-means centers. First run a K-means analysis with cluster centers from the calibration sample (derived from mba_car_pref1.sav in Example 8.2). Second run a K-means analysis with cluster centers from the calibration sample—identically repeating steps of Example 8.2.

Output 8.3.A: *Results from assigning observations in validation sample to closest cluster centroid from calibration sample*

Final Cluster Centers

	Cluster				
	1	2	3	4	5
COL2	4	7	8	7	7
COL3	6	6	5	5	3
COL4	4	5	2	3	5
COL5	7	6	6	5	4
COL6	5	6	3	4	6
COL7	3	3	1	1	2
COL8	4	7	5	6	7
COL9	5	6	7	4	5
COL10	3	7	4	8	6
COL11	5	6	3	2	3

Number of Cases in each Cluster

Cluster	1	17.000
	2	35.000
	3	17.000
	4	33.000
	5	45.000
Valid		147.000
Missing		4.000

Distances between Final Cluster Centers

Cluster	1	2	3	4	5
1		6.316	6.142	7.754	7.290
2	6.316		6.327	6.059	4.824
3	6.142	6.327		5.711	6.641
4	7.754	6.059	5.711		5.267
5	7.290	4.824	6.641	5.267	

Output 8.3.B: *Cross-tabulation of cluster results from validation sample*

The column variable, QCL_1, tabulates cluster membership within the validation sample based on centroids generated from Wards method clustering in the calibration sample. The row variable, QCL_2, tabulates cluster membership within the validation sample based on centroids generated from Wards method clustering in the validation sample. The match up is generally satisfactory. For example, Cluster 1 of QCL_1 and Cluster 5 of QCL_2 have a 94% overlap. Cluster 4 does not match well in either QCL_1 and QCL_2.

QCL_2 * QCL_1 Crosstabulation

			QCL_1					
			1	2	3	4	5	Total
QCL_2	1	Count	1	2	1	26	1	31
		% within QCL_2	3.2%	6.5%	3.2%	83.9%	3.2%	100.0%
	2	Count	1	0	0	5	31	37
		% within QCL_2	2.7%	.0%	.0%	13.5%	83.8%	100.0%
	3	Count	0	1	15	2	0	18
		% within QCL_2	.0%	5.6%	83.3%	11.1%	.0%	100.0%
	4	Count	0	32	0	0	13	45
		% within QCL_2	.0%	71.1%	.0%	.0%	28.9%	100.0%
	5	Count	15	0	1	0	0	16
		% within QCL_2	93.8%	.0%	6.3%	.0%	.0%	100.0%
Total		Count	17	35	17	33	45	147
		% within QCL_2	11.6%	23.8%	11.6%	22.4%	30.6%	100.0%

9 Canonical Correlation Using SPSS

Example 9.1: *Canonical correlation of IRI Factbook data*

To run the canonical correlation macro open the active data file for this example, factbook.sav. Rename variables col1 through col11 (see list of variable names in syntax viewer sample below). Open a new syntax file and type in the following code:

```
INCLUDE file 'C:\Program Files\SPSS\canonical correlation.sps'.
CANCORR set1=
/set2=.
```

Define variable sets 1 and 2 by listing variables after each equal sign. Verify the correct path of Canonical Correlation.SPS, highlight all syntax, and run.

The INCLUDE keyword directs SPSS to run all commands in the designated syntax file. SPSS syntax file Canonical Correlation.SPS contains macro commands for running canonical correlation analysis. Canonical Correlation.SPS is included within the Base SPSS package and can typically be found in the main SPSS directory with path 'C:\Program Files\SPSS.' The term 'CANCORR' is not an SPSS Base syntax keyword, so no Help information or syntax reference is available. 'CANCORR' merely calls the macro language of Canonical Correlation.SPS.

```
9.1macro.SPS - SPSS Syntax Editor
File Edit View Analyze Graphs Utilities Run Window Help

*9.1: Open factbook.sav as working data file.

RENAME VARIABLES (col1 to col11 = cat_name penet purhh pcycle price pvtsh feat disp pcut scoup mcoup).

*Run canonical correlation macro (modify the first command line to include thecorrect file path).

include file 'C:\Program Files\SPSS\canonical correlation.sps'.
cancorr set1= pcycle penet price purhh pvtsh
/set2=disp feat mcoup pcut scoup.

*Alternate approach: MANOVA (to run, delete the asterisk before the MANOVA keyword).

*MANOVA pcycle penet price purhh pvtsh with disp feat mcoup pcut scoup
/DISCRIM all alpha(1)
/PRINT=sig(eigen dim multiv) param (estim)
/DESIGN
```

Canonical correlation analyses can alternatively be performed with the SPSS MANOVA command. MANOVA command syntax cannot be created from menus, it must be typed into a syntax file. One limitation of the MANOVA command for canonical correlation analysis is that it has no option to save canonical scores.

MANOVA has the basic form:

> MANOVA *variable list* 1 WITH *variable list* 2
> /DISCRIM *options*
> /PRINT *options*.

Output 9.1: *Canonical Correlation results on IRI Factbook Data*

*The SPSS canonical correlation macro produces considerable output, including raw and standardized canonical variates, redundancy analysis and saves canonical scores to the working data file. Optional output lists canonical equations. Only a portion of total canonical correlation output is included here.

```
Correlations for Set-1
          PCYCLE    PENET     PRICE     PURHH     PVTSH
PCYCLE    1.0000    -.4779    -.1457    -.7189    -.1267
 PENET    -.4779    1.0000    -.2218     .6175     .4091
 PRICE    -.1457    -.2218    1.0000     .0680    -.2802
 PURHH    -.7189     .6175     .0680    1.0000     .2463
 PVTSH    -.1267     .4091    -.2802     .2463    1.0000


Correlations for Set-2
          DISP      FEAT      MCOUP     PCUT      SCOUP
 DISP     1.0000     .5351    -.0376     .5152     .3746
 FEAT      .5351    1.0000    -.0435     .9175     .6739
MCOUP     -.0376    -.0435    1.0000    -.0395    -.0652
 PCUT      .5152     .9175    -.0395    1.0000     .5884
SCOUP      .3746     .6739    -.0652     .5884    1.0000


Correlations Between Set-1 and Set-2
          DISP      FEAT      MCOUP     PCUT      SCOUP
PCYCLE    -.2523    -.3790     .0494    -.3945    -.1780
 PENET     .4607     .5799     .0526     .5695     .3887
 PRICE    -.1107    -.0013     .2367    -.1082     .0740
 PURHH     .2126     .3727    -.0263     .3677     .2609
 PVTSH     .1318     .2697    -.2149     .2953     .2228

Canonical Correlations
1      .642
2      .483
3      .265
4      .114
5      .032
Test that remaining correlations are zero:
      Wilk's    Chi-SQ       DF       Sig.
1      .413    287.260    25.000     .000
2      .703    114.552    16.000     .000
3      .917     28.182     9.000     .001
4      .986      4.574     4.000     .334
5      .999       .335     1.000     .563

Canonical Loadings for Set-1
              1          2          3          4          5
PCYCLE     -.582       .320      -.060      -.697       .263
 PENET      .956      -.114       .042      -.223      -.145
 PRICE     -.011       .769       .285       .569       .059
```

	1	2	3	4	5
PURHH	.555	-.148	.389	.207	-.690
PVTSH	.336	-.465	.705	-.245	.337

Cross Loadings for Set-1

	1	2	3	4	5
PCYCLE	-.374	.155	-.016	-.079	.008
PENET	.614	-.055	.011	-.025	-.005
PRICE	-.007	.372	.075	.065	.002
PURHH	.356	-.071	.103	.024	-.022
PVTSH	.216	-.225	.187	-.028	.011

Canonical Loadings for Set-2

	1	2	3	4	5
DISP	.730	-.136	-.384	-.412	.362
FEAT	.939	-.073	.293	.157	.046
MCOUP	.156	.717	-.427	.069	-.523
PCUT	.896	-.321	.184	.063	-.238
SCOUP	.617	.167	.614	-.462	-.024

Cross Loadings for Set-2

	1	2	3	4	5
DISP	.469	-.066	-.102	-.047	.012
FEAT	.603	-.035	.078	.018	.001
MCOUP	.100	.347	-.113	.008	-.017
PCUT	.575	-.155	.049	.007	-.008
SCOUP	.397	.081	.163	-.053	-.001

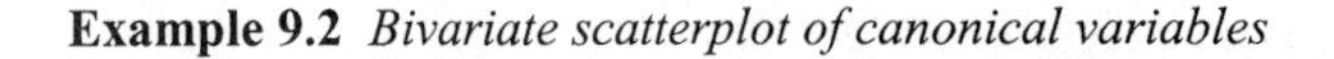

Example 9.2 *Bivariate scatterplot of canonical variables*

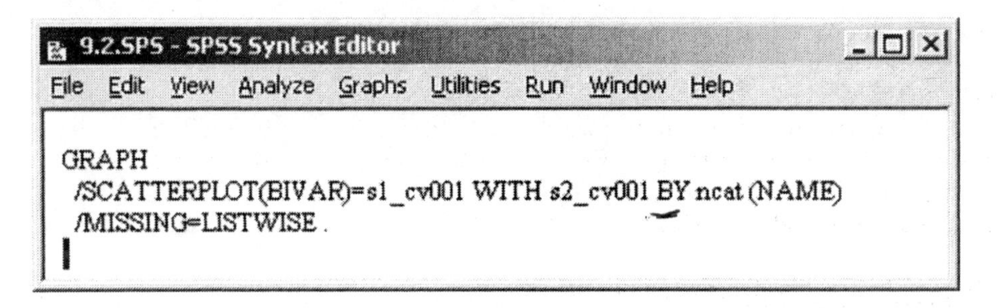

9.2.SPS - SPSS Syntax Editor

File Edit View Analyze Graphs Utilities Run Window Help

```
GRAPH
 /SCATTERPLOT(BIVAR)=s1_cv001 WITH s2_cv001 BY ncat (NAME)
 /MISSING=LISTWISE .
```

Output 9.2 *Bivariate scatterplot of canonical variables*

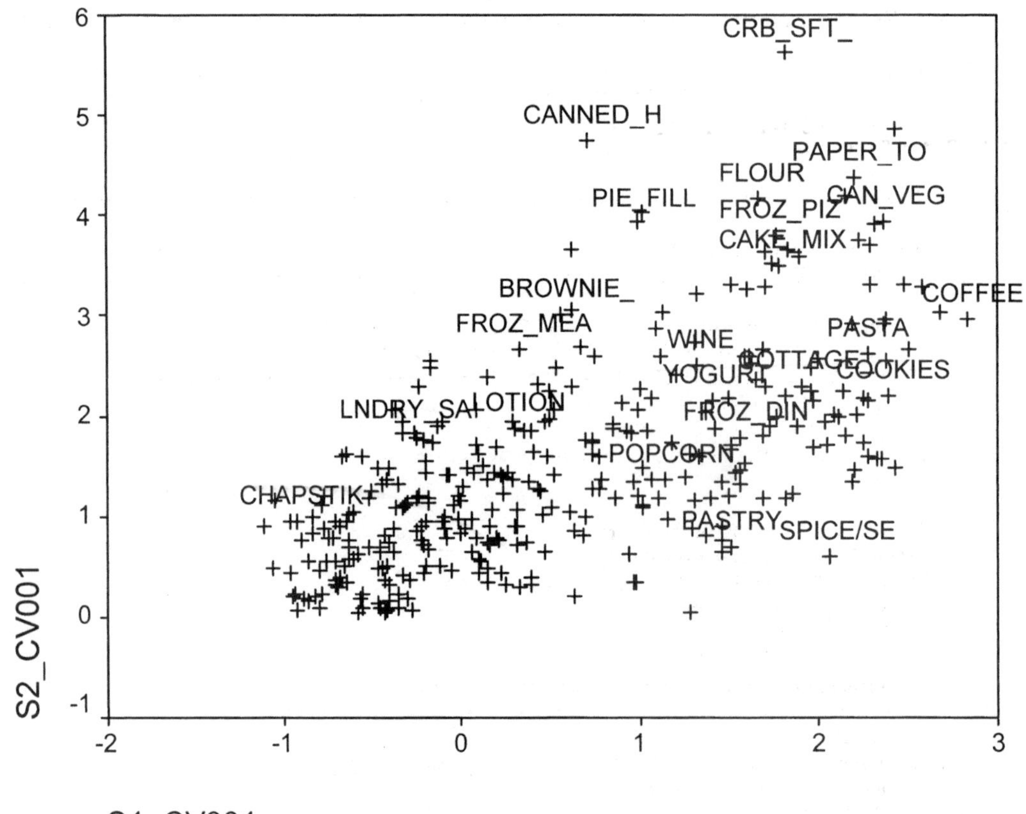

Example 9.3: *Split sample validation of canonical correlation analysis of IRI Factbook data*

Example 9.3 is a split sample validation of similar design the to Example 8.2 and 8.3. Begin by opening Factbook_1.sav as the working data file. The first programming step runs the CANCORR macro, identically to Example 9.1 syntax. For the second step use the coefficients obtained from this validation sample to compute canonical variables Struct1, Struct2, Struct3, Promo1, Promo2 and Promo3. Validation coefficients are listed in Output 9.3.B. Run canonical correlation analysis on the calibration sample for the third programming step. Finally create bivariate correlation tables estimated from validation and calibration canonical score coefficients.

9.3.SPS - SPSS Syntax Editor

File Edit View Analyze Graphs Utilities Run Window Help

```
*9.3: Split sample validation of canonical correlation analysis of IRI Factbook data.
*9.3.A: Validation sample. Open factbook_1.sav as working data file.

RENAME VARIABLES (col1 to col11 = cat_name penet purhh pcycle price pvtsh feat disp pcut scoup mcoup).

*Run canonical correlation macro (modify file path for your PC).
INCLUDE FILE 'C:\Program Files\SPSS\canonical correlation.sps'.
CANCORR set1= pcycle penet price purhh pvtsh
/set2=disp feat mcoup pcut scoup.

*9.3.B: Use coefficients obtained from validation sample to compute canonical variables with validation data.

*Open factbook_2.sav as working data file.
RENAME VARIABLES (col1 to col11 = cat_name penet purhh pcycle price pvtsh feat disp pcut scoup mcoup).

COMPUTE Struct1=(.037*penet)-(.01*PCYCLE) +(.37*price)-(.09*purhh)+(.006*pvtsh).
COMPUTE Struct2=(.015 *penet)+(.026 *PCYCLE) +(.762*price)-(.053*purhh)-(.023*pvtsh).
COMPUTE Struct3=(.021 *penet)+(.024 *PCYCLE) + (.522*price)+(.206*purhh)+(.06*pvtsh).
COMPUTE promo1=(.02*mcoup)+(.029*disp) +(.145*feat) -(.049*scoup)-(.033*pcut).
COMPUTE promo2=(.079*mcoup)-(.036*disp) +(.074*feat) +(.604*scoup)-(.134*pcut).
COMPUTE promo3=(.01*feat) +(.677*scoup)-(.028*pcut)-(.068*mcoup)-(.077*disp).
EXECUTE.

*9.3.C: Calibration data canonical correlation.
*Run canonical correlation macro (modify file path for your PC).
INCLUDE FILE 'C:\Program Files\SPSS\canonical correlation.sps'.
CANCORR set1= pcycle penet price purhh pvtsh
/set2=disp feat mcoup pcut scoup.

*9.3.D: Bivariate correlation matrix of based on score coefficients.
*estimated from validation data (STRUCT, PROMO) and from calibration data (S1,S2).
CORRELATIONS
 /VARIABLES=s1_cv001 s1_cv002 s1_cv003 s2_cv001 s2_cv002 s2_cv003 struct1
 struct2 struct3 promo1 promo2 promo3
 /MISSING=PAIRWISE .
```

Output 9.3.A: *Canonical correlation analysis of validation sample of IRI Factbook data*

Run MATRIX procedure:

Correlations for Set-1

	PCYCLE	PENET	PRICE	PURHH	PVTSH
PCYCLE	1.0000	-.4422	-.1210	-.7046	-.0832
PENET	-.4422	1.0000	-.2100	.6876	.3832
PRICE	-.1210	-.2100	1.0000	-.0364	-.3182
PURHH	-.7046	.6876	-.0364	1.0000	.2322
PVTSH	-.0832	.3832	-.3182	.2322	1.0000

Correlations for Set-2

	DISP	FEAT	MCOUP	PCUT	SCOUP
DISP	1.0000	.4703	-.0915	.4894	.3245
FEAT	.4703	1.0000	-.0720	.9379	.7083
MCOUP	-.0915	-.0720	1.0000	-.0454	-.1056
PCUT	.4894	.9379	-.0454	1.0000	.6031
SCOUP	.3245	.7083	-.1056	.6031	1.0000

Correlations Between Set-1 and Set-2

	DISP	FEAT	MCOUP	PCUT	SCOUP
PCYCLE	-.2469	-.3497	.0628	-.3533	-.1272
PENET	.4102	.5877	.0073	.5703	.3640
PRICE	-.0967	.0941	.1859	.0181	.1918
PURHH	.2582	.4070	-.0882	.4120	.2291
PVTSH	.1664	.2707	-.2746	.2636	.1940

Canonical Correlations

1	.662
2	.432
3	.291
4	.098
5	.066

Test that remaining correlations are zero:

	Wilk's	Chi-SQ	DF	Sig.
1	.412	141.393	25.000	.000
2	.734	49.296	16.000	.000
3	.903	16.355	9.000	.060
4	.986	2.243	4.000	.691
5	.996	.706	1.000	.401

Canonical Loadings for Set-1

	1	2	3	4	5
PCYCLE	-.526	.434	-.156	-.516	.494
PENET	.910	-.246	.099	-.317	-.010
PRICE	.165	.754	-.274	.476	-.321
PURHH	.592	-.392	-.111	-.178	-.672
PVTSH	.333	-.538	-.669	-.058	.387

Canonical Loadings for Set-2

	1	2	3	4	5
DISP	.590	-.336	.302	-.324	.585
FEAT	.968	-.093	-.199	.004	-.122
MCOUP	.098	.697	.661	-.089	-.245
PCUT	.894	-.251	-.075	-.218	-.292
SCOUP	.663	.291	-.564	-.373	.137

Output 9.3.B: *Validation canonical coefficients which are used with calibration data.*

Raw canonical coefficients for DEPENDENT variables (Set-1)

	1	2	3	4	5
PCYCLE	-.010	.026	-.024	-.056	-.005
PENET	.037	.015	.021	-.009	.023
PRICE	.370	.762	-.522	.312	-.115
PURHH	-.090	-.053	-.206	-.268	-.393
PVTSH	.006	-.023	-.060	.016	.024

Output 9.3.C: *Canonical correlation analysis of calibration sample of IRI Factbook data*

Canonical Correlations

1	.655
2	.567
3	.307
4	.143
5	.038

Test that remaining correlations are zero:

	Wilk's	Chi-SQ	DF	Sig.
1	.343	169.577	25.000	.000
2	.601	80.595	16.000	.000
3	.886	19.143	9.000	.024
4	.978	3.483	4.000	.480
5	.999	.229	1.000	.632

Canonical Loadings for Set-1

	1	2	3	4	5
PCYCLE	-.642	-.123	.197	-.663	.306
PENET	.950	-.191	-.172	-.179	-.003
PRICE	-.324	-.639	-.312	.620	.075
PURHH	.519	-.011	-.678	.189	-.484
PVTSH	.444	.376	-.547	-.108	.592

Canonical Loadings for Set-2

	1	2	3	4	5
DISP	.773	-.286	.368	-.348	-.253
FEAT	.903	-.114	-.244	.254	-.216
MCOUP	.010	-.751	-.035	.100	.651
PCUT	.951	.181	-.136	.182	.107
SCOUP	.638	-.074	-.612	-.456	-.070

Output 9.3.D: *Correlations between canonical variates estimated from calibration data and canonical variates estimated from validation data*

Output 9.3.D is a table of correlations between canonical variates based on score coefficients estimated from calibration data (S1, S2) and canonical variates based on score coefficients estimated from validation data (STRUCT, PROMO).

		STRUCT1	STRUCT2	STRUCT3	PROMO1	PROMO2	PROMO3
S1_CV001	Pearson Correlation	0.88	-0.50	0.58	0.58	-0.26	-0.20
	Sig. (2-tailed)	0.00	0.00	0.00	0.00	0.00	0.01
S1_CV002	Pearson Correlation	-0.39	-0.83	-0.15	-0.21	-0.43	0.36
	Sig. (2-tailed)	0.00	0.00	0.05	0.01	0.00	0.00
S1_CV003	Pearson Correlation	-0.11	0.14	-0.79	-0.04	-0.13	-0.18
	Sig. (2-tailed)	0.17	0.08	0.00	0.61	0.11	0.02
S2_CV001	Pearson Correlation	0.58	-0.33	0.38	0.88	-0.40	-0.30
	Sig. (2-tailed)	0.00	0.00	0.00	0.00	0.00	0.00
S2_CV002	Pearson Correlation	-0.22	-0.47	-0.09	-0.37	-0.77	0.64
	Sig. (2-tailed)	0.00	0.00	0.26	0.00	0.00	0.00
S2_CV003	Pearson Correlation	-0.03	0.04	-0.24	-0.13	-0.41	-0.57
	Sig. (2-tailed)	0.68	0.59	0.00	0.09	0.00	0.00

Output 9.3.E: *Correlations among canonical variates based on score coefficients estimated from validation data*

		STRUCT1	STRUCT2	STRUCT3	PROMO1	PROMO2	PROMO3
STRUCT1	Pearson Correlation	1.00	-0.09	0.68	0.60	-0.04	-0.30
	Sig. (2-tailed)	.	0.25	0.00	0.00	0.58	0.00
STRUCT2	Pearson Correlation	-0.09	1.00	-0.25	-0.12	0.48	-0.23
	Sig. (2-tailed)	0.25	.	0.00	0.12	0.00	0.00
STRUCT3	Pearson Correlation	0.68	-0.25	1.00	0.40	0.02	-0.03
	Sig. (2-tailed)	0.00	0.00	.	0.00	0.83	0.66
PROMO1	Pearson Correlation	0.60	-0.12	0.40	1.00	-0.08	-0.38
	Sig. (2-tailed)	0.00	0.12	0.00	.	0.28	0.00
PROMO2	Pearson Correlation	-0.04	0.48	0.02	-0.08	1.00	-0.23
	Sig. (2-tailed)	0.58	0.00	0.83	0.28	.	0.00
PROMO3	Pearson Correlation	-0.30	-0.23	-0.03	-0.38	-0.23	1.00
	Sig. (2-tailed)	0.00	0.00	0.66	0.00	0.00	.

10 Structural Equation Models with Latent Variables Using Amos

10.1 Amos Basic Editor
10.2 Structural equation model of Adoption of Innovation
10.3 Structural equation model of Service Quality --

This workbook chapter demonstrates structural equation models with Amos Basic syntax. Users may also use the Amos graphical programming interface demonstrated in Chapter 6.

Example 10.1 *The Amos Basic Editor*
Open the Amos Basic editor by double-clicking its icon. A working Amos program can be created from a few basic command prefixes, as the following illustration, taken from text Chapter 10, demonstrates.

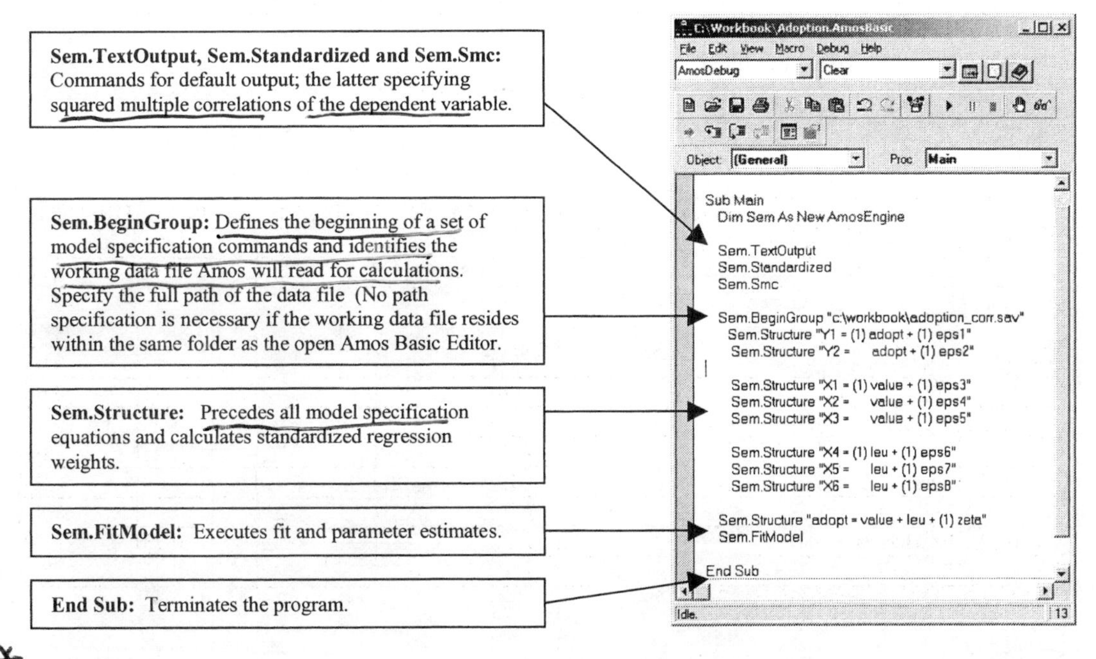

Single quotes (') define comment lines. Create syntax by text-editing within the Amos Basic Editor, or copy-paste from a word processor.

Example 10.2 *Adoption of Technological Innovation*

Example 10.2 demonstrates the adoption of innovation model discussed in Chapter 10 of the text book (see p. 367). The working data file is adoption_corr.sav, obtained from the CD-ROM attached to this guide. Alternatively modify adoption.sav, obtained from the Chapter 10 folder in the Analyzing Multivariate Data CD-ROM. The following syntax converts adoption.sav into correlation matrix format.

create matrix.SPS - SPSS Syn...

File Edit View Analyze Graphs Utilities Run Window Help

```
*Open adoption.sav from the file icon.
*Create a correlation matrix.

CORRELATIONS
 /VARIABLES=y1 y2 x1 x2 x3 x4 x5 x6
 /MISSING=PAIRWISE
/matrix=out (*).

SAVE OUTFILE= 'C:\adoption_corr.sav'.
```

The adoption_corr.sav matrix file contains means, standard deviations, number of observations and correlation coefficients. The automatically created variable rowtype_ identifies and distinguishes these data types. Variables col1 through col5 represent expert ratings.

adoption_corr.sav - SPSS Data Editor

File Edit View Data Transform Analyze Graphs Utilities Window Help

1 : rowtype_ MEAN

	rowtype_	varname_	y1	y2	x1	x2	x3	x4	x5	x6
1	MEAN		15.251	31.311	3.604	3.207	3.340	3.668	3.473	3.766
2	STDDEV		27.061	31.048	.997	.916	.954	.871	.973	.806
3	N	Y1	171.000	167.000	170.000	171.000	171.000	170.000	169.000	170.000
4	N	Y2	167.000	180.000	179.000	180.000	180.000	180.000	178.000	180.000
5	N	X1	170.000	179.000	187.000	187.000	187.000	183.000	181.000	183.000
6	N	X2	171.000	180.000	187.000	188.000	188.000	184.000	182.000	184.000
7	N	X3	171.000	180.000	187.000	188.000	188.000	184.000	182.000	184.000
8	N	X4	170.000	180.000	183.000	184.000	184.000	184.000	182.000	184.000
9	N	X5	169.000	178.000	181.000	182.000	182.000	182.000	182.000	182.000
10	N	X6	170.000	180.000	183.000	184.000	184.000	184.000	182.000	184.000
11	CORR	Y1	1.000	.599	.263	.248	.222	.478	.464	.360
12	CORR	Y2	.599	1.000	.110	.156	.104	.571	.580	.481
13	CORR	X1	.263	.110	1.000	.437	.542	.187	.208	.122
14	CORR	X2	.248	.156	.437	1.000	.421	.124	.139	.051
15	CORR	X3	.222	.104	.542	.421	1.000	.155	.191	.054
16	CORR	X4	.478	.571	.187	.124	.155	1.000	.763	.628
17	CORR	X5	.464	.580	.208	.139	.191	.763	1.000	.720
18	CORR	X6	.360	.481	.122	.051	.054	.628	.720	1.000

Data View | Variable View

SPSS Processor is ready

The model has one dependent construct ("adopt," measured by Y_1 and Y_2). The model has two independent constructs ("value," measured by X_1, X_2, and X_3) and ("leu," measured by X_4, X_5, and X_6). To run Amos open the **Macro** menu and click on **Run**. The syntax below creates approximately 25 tables of output. A Variable Summary table classifies variables as endogenous, exogenous, observed, unobserved.

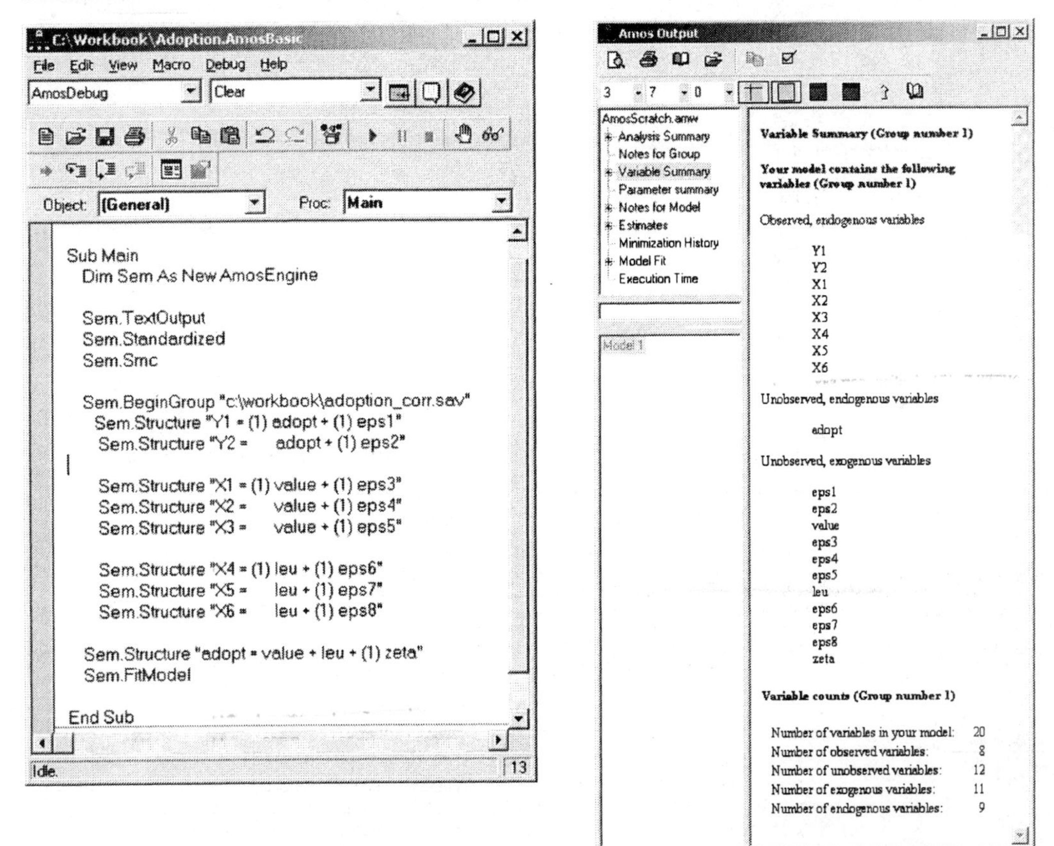

The Results table is an initial test for adequacy of model fit. A probability level of .05 or less suggests data does not fit the model well. Six Estimates tables contain maximum likelihood estimates for regression weights, standardized regression weights, covariances, correlations, variances, and squared multiple correlations.

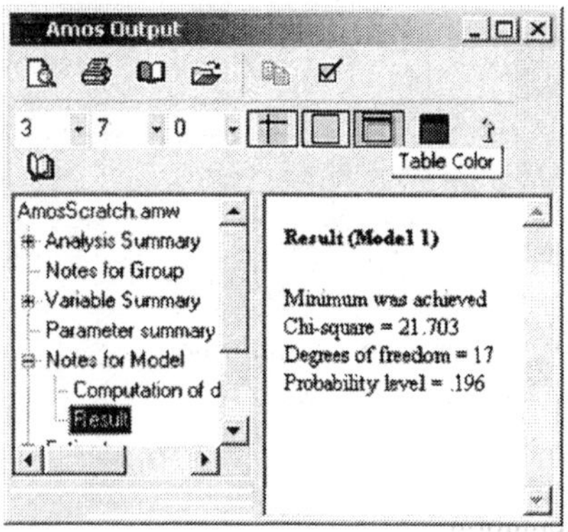

Result (Model 1)

Minimum was achieved
Chi-square = 21.703
Degrees of freedom = 17
Probability level = .196

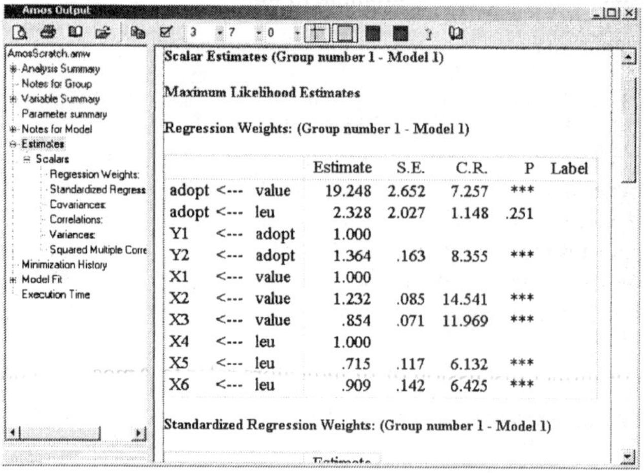

Scalar Estimates (Group number 1 - Model 1)

Maximum Likelihood Estimates

Regression Weights: (Group number 1 - Model 1)

			Estimate	S.E.	C.R.	P	Label
adopt	<---	value	19.248	2.652	7.257	***	
adopt	<---	leu	2.328	2.027	1.148	.251	
Y1	<---	adopt	1.000				
Y2	<---	adopt	1.364	.163	8.355	***	
X1	<---	value	1.000				
X2	<---	value	1.232	.085	14.541	***	
X3	<---	value	.854	.071	11.969	***	
X4	<---	leu	1.000				
X5	<---	leu	.715	.117	6.132	***	
X6	<---	leu	.909	.142	6.425	***	

Standardized Regression Weights: (Group number 1 - Model 1)

The Standardized Regression Weights table indicates that when Value goes up by 1, Adopt also rises by .728 standard deviations. The Squared Multiple Correlations table indicates that predictors of adopt explain 57.3 percent of its variance. Therefore, the error variance of Adopt is approximately 42.7 percent of its total variance.

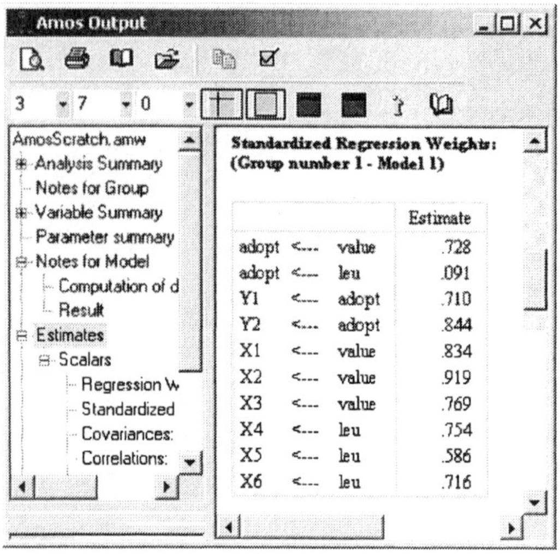

Standardized Regression Weights: (Group number 1 - Model 1)

			Estimate
adopt	<---	value	.728
adopt	<---	leu	.091
Y1	<---	adopt	.710
Y2	<---	adopt	.844
X1	<---	value	.834
X2	<---	value	.919
X3	<---	value	.769
X4	<---	leu	.754
X5	<---	leu	.586
X6	<---	leu	.716

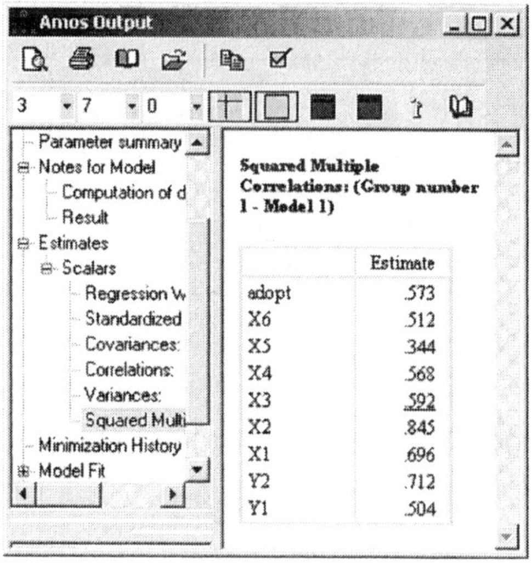

Squared Multiple Correlations: (Group number 1 - Model 1)

	Estimate
adopt	.573
X6	.512
X5	.344
X4	.568
X3	.592
X2	.845
X1	.696
Y2	.712
Y1	.504

Model Fit Summary

Amos output displays approximately two dozen indicators for assessing model fit, many of which are the subject of ongoing debate. Model fit output is displayed for three types of model: the user-specified model, a saturated model, and an independence model. The latter two may be considered rough extremes on a continuum of model fit. The saturated model by definition fits any set of data perfectly. The independence model is so constrained it fits virtually no data.

Two commonly used fit indicators based on minimum sample discrepancy are reported in the CMIN table: P and CMIN/DF. The P value can be used to test whether a model is discrepant from the population. The Example 10.2 user-specified Adoption of Innovation model P of .196 indicates the model fits the population. The CMIN/DF indicator is the minimum discrepancy divided by degrees of freedom. CMIN/DF fits well if it is close to 1. Some sources suggest a CMIN/DF score of less the 5 indicates reasonable fit. The CMIN/DF of 1.277 therefore provides further evidence that the Adoption of Innovation model has adequate fit to data. FMIN, or Minimum Discrepancy Function, is yet another minimum sample discrepancy-based indicator of model fit.

For further discussion of fit indicators refer to Amos Help documentation.

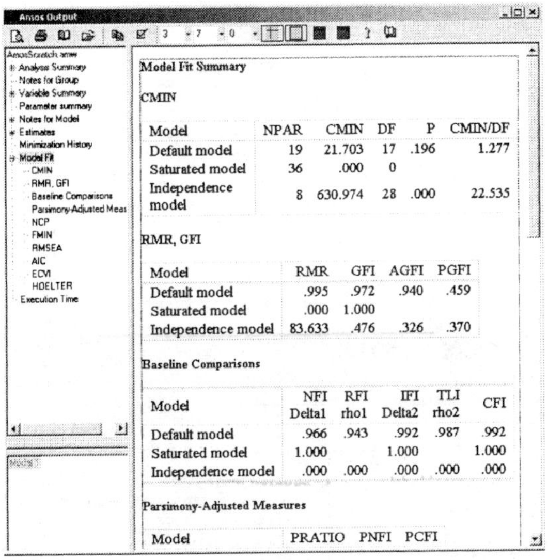

Model Fit Summary

CMIN

Model	NPAR	CMIN	DF	P	CMIN/DF
Default model	19	21.703	17	.196	1.277
Saturated model	36	.000	0		
Independence model	8	630.974	28	.000	22.535

RMR, GFI

Model	RMR	GFI	AGFI	PGFI
Default model	.995	.972	.940	.459
Saturated model	.000	1.000		
Independence model	83.633	.476	.326	.370

Baseline Comparisons

Model	NFI Delta1	RFI rho1	IFI Delta2	TLI rho2	CFI
Default model	.966	.943	.992	.987	.992
Saturated model	1.000		1.000		1.000
Independence model	.000	.000	.000	.000	.000

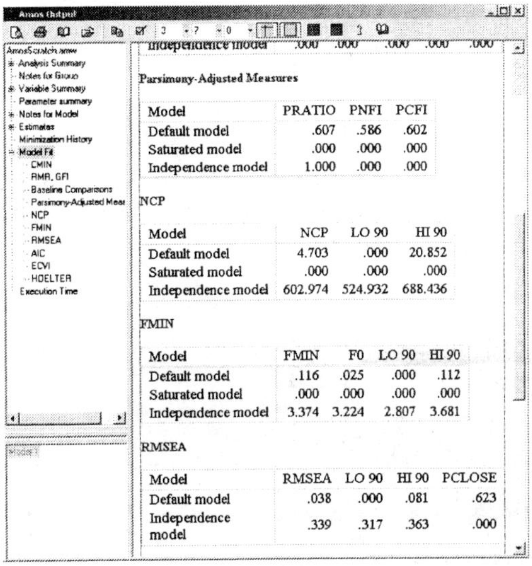

Parsimony-Adjusted Measures

Model	PRATIO	PNFI	PCFI
Default model	.607	.586	.602
Saturated model	.000	.000	.000
Independence model	1.000	.000	.000

NCP

Model	NCP	LO 90	HI 90
Default model	4.703	.000	20.852
Saturated model	.000	.000	.000
Independence model	602.974	524.932	688.436

FMIN

Model	FMIN	F0	LO 90	HI 90
Default model	.116	.025	.000	.112
Saturated model	.000	.000	.000	.000
Independence model	3.374	3.224	2.807	3.681

RMSEA

Model	RMSEA	LO 90	HI 90	PCLOSE
Default model	.038	.000	.081	.623
Independence model	.339	.317	.363	.000

Amos hypertext Help

Amos help documentation is hypertext embedded in output table headings and values.

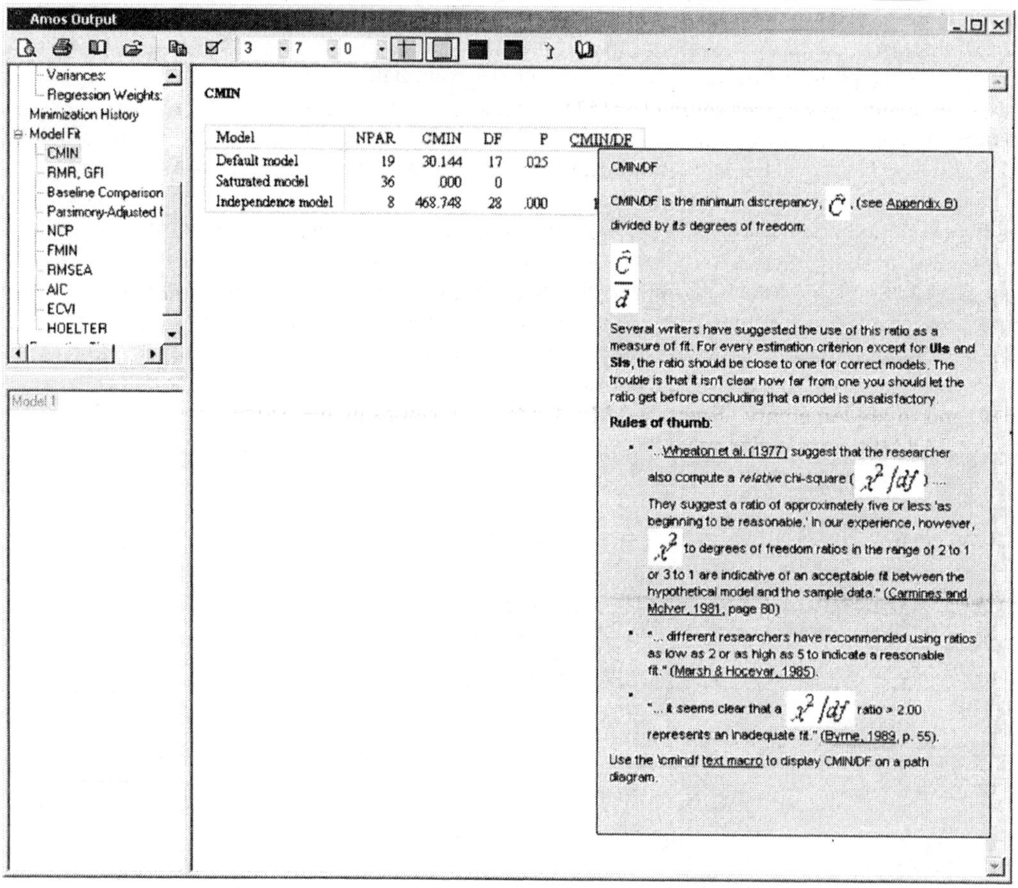

Amos Output

3 7 0

Variances:
Regression Weights:
Minimization History
Model Fit
CMIN
RMR, GFI
Baseline Comparison
Parsimony-Adjusted I
NCP
FMIN
RMSEA
AIC
ECVI
HOELTER

Model 1

CMIN

Model	NPAR	CMIN	DF	P	CMIN/DF
Default model	19	30.144	17	.025	
Saturated model	36	.000	0		
Independence model	8	468.748	28	.000	1

CMIN/DF

CMIN/DF is the minimum discrepancy, $\hat{C}$, (see Appendix B) divided by its degrees of freedom.

$$\frac{\hat{C}}{d}$$

Several writers have suggested the use of this ratio as a measure of fit. For every estimation criterion except for **Uls** and **Sls**, the ratio should be close to one for correct models. The trouble is that it isn't clear how far from one you should let the ratio get before concluding that a model is unsatisfactory.

Rules of thumb:

- "...Wheaton et al. (1977) suggest that the researcher also compute a *relative* chi-square (χ^2/df) .... They suggest a ratio of approximately five or less 'as beginning to be reasonable.' In our experience, however, χ^2 to degrees of freedom ratios in the range of 2 to 1 or 3 to 1 are indicative of an acceptable fit between the hypothetical model and the sample data." (Carmines and McIver, 1981, page 80)
- "... different researchers have recommended using ratios as low as 2 or as high as 5 to indicate a reasonable fit." (Marsh & Hocevar, 1985).
- "... it seems clear that a χ^2/df ratio > 2.00 represents an inadequate fit." (Byrne, 1989, p. 55).

Use the \cmindf text macro to display CMIN/DF on a path diagram.

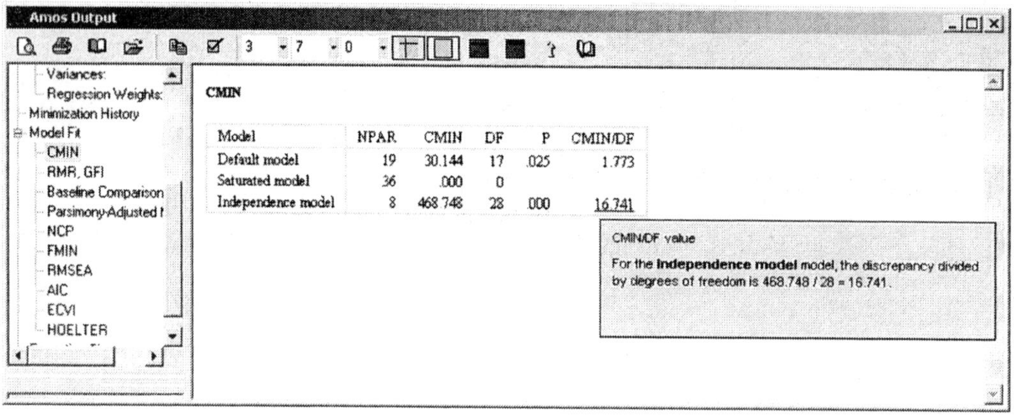

Amos Output

3 7 0

Variances:
Regression Weights:
Minimization History
Model Fit
CMIN
RMR, GFI
Baseline Comparison
Parsimony-Adjusted I
NCP
FMIN
RMSEA
AIC
ECVI
HOELTER

CMIN

Model	NPAR	CMIN	DF	P	CMIN/DF
Default model	19	30.144	17	.025	1.773
Saturated model	36	.000	0		
Independence model	8	468.748	28	.000	16.741

CMIN/DF value

For the **Independence model** model, the discrepancy divided by degrees of freedom is 468.748 / 28 = 16.741.

Example 10.3 *Structural Equation Model of Service Quality*

Example 10.3 demonstrates an Amos solution for practice exercise 10.2 in the textbook Analyzing Multivariate Data. Discussion of syntax and output is based on James Lattin's problem solution for analyzing Multivariate Data Exercise 10.2.

Data is a correlation matrix of data collected for Crosby, Evans and Cowles' 1990 study on impact of relationship quality on services selling (n=151). The working data file is serv_qual.sav, obtained from the CD-ROM attached to this guide. Alternatively modify service_quality.sav, obtained from the Chapter 9 folder in the Analyzing Multivariate Data CD-ROM.

To modify service_quality.sav, in the SPSS Data Editor **Variable View** add variables Rowtype_ and Varname_ and delete variable Col1. Rename variables Col2-Col11 to be Y1,Y2, Y3,Y4,X1,X2,X3,X4,X5,X6. In the SPSS Data Editor **Data View** add data type designations to the rowtype_ column. Data type designations include 'n' (sample size values), 'stddev,' (standard deviation values), and 'corr' for each matrix row. Add variable names to the varname_ column for each row corresponding to a row in the correlation matrix. Note that the rows corresponding to 'stddev' and 'n'are left empty. Enter variable standard deviations in the 'stddev' row (row 1 in this example). Add the sample size values to the 'n' row (here the value 151 is entered for all cells in the row).

The complete data file should appear as illustrated here :

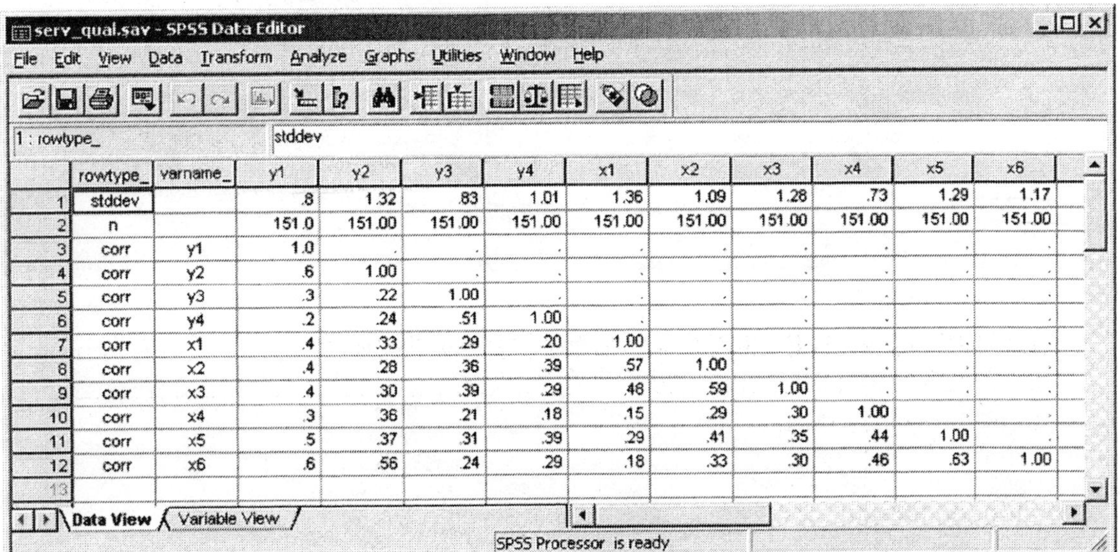

serv_qual.sav - SPSS Data Editor

File Edit View Data Transform Analyze Graphs Utilities Window Help

1 : rowtype_ stddev

	rowtype_	varname_	y1	y2	y3	y4	x1	x2	x3	x4	x5	x6
1	stddev		.8	1.32	.83	1.01	1.36	1.09	1.28	.73	1.29	1.17
2	n		151.0	151.00	151.00	151.00	151.00	151.00	151.00	151.00	151.00	151.00
3	corr	y1	1.0	.	.	.	.	.	.	.	.	.
4	corr	y2	.6	1.00	.	.	.	.	.	.	.	.
5	corr	y3	.3	.22	1.00	.	.	.	.	.	.	.
6	corr	y4	.2	.24	.51	1.00	.	.	.	.	.	.
7	corr	x1	.4	.33	.29	.20	1.00	.	.	.	.	.
8	corr	x2	.4	.28	.36	.39	.57	1.00	.	.	.	.
9	corr	x3	.4	.30	.39	.29	.48	.59	1.00	.	.	.
10	corr	x4	.3	.36	.21	.18	.15	.29	.30	1.00	.	.
11	corr	x5	.5	.37	.31	.39	.29	.41	.35	.44	1.00	.
12	corr	x6	.6	.56	.24	.29	.18	.33	.30	.46	.63	1.00
13												

Data View Variable View

SPSS Processor is ready

The dependent construct is "attitude," measured by Y_1 and Y_2. The two independent constructs are "similarity" and "interaction," the first measured by X_1, X_2, and X_3, the second measured by X_4, X_5, and X_6. 17 degrees of freedom are required to calculate the model.

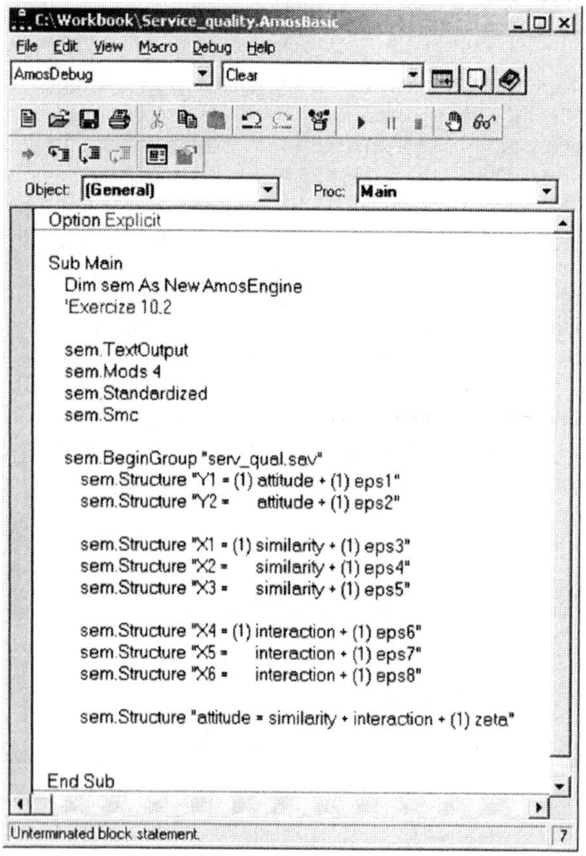

C:\Workbook\Service_quality.AmosBasic

File Edit View Macro Debug Help

AmosDebug | Clear

Object: (General) Proc: Main

```
Option Explicit

Sub Main
  Dim sem As New AmosEngine
  'Exercize 10.2

  sem.TextOutput
  sem.Mods 4
  sem.Standardized
  sem.Smc

  sem.BeginGroup "serv_qual.sav"
    sem.Structure "Y1 = (1) attitude + (1) eps1"
    sem.Structure "Y2 =     attitude + (1) eps2"

    sem.Structure "X1 = (1) similarity + (1) eps3"
    sem.Structure "X2 =     similarity + (1) eps4"
    sem.Structure "X3 =     similarity + (1) eps5"

    sem.Structure "X4 = (1) interaction + (1) eps6"
    sem.Structure "X5 =     interaction + (1) eps7"
    sem.Structure "X6 =     interaction + (1) eps8"

    sem.Structure "attitude = similarity + interaction + (1) zeta"

End Sub
```

Unterminated block statement. 7

Summary of models

Model fit tests provide conflicting indications. The chi-square test is significant at the p=0.025 level, which would typically suggest rejecting the model. The P of .025 is also significant suggesting rejection of the model. On the other hand, the CMIN/DF (1.773), GFI (0.95) and AGFI (0.90) all suggest good fit.

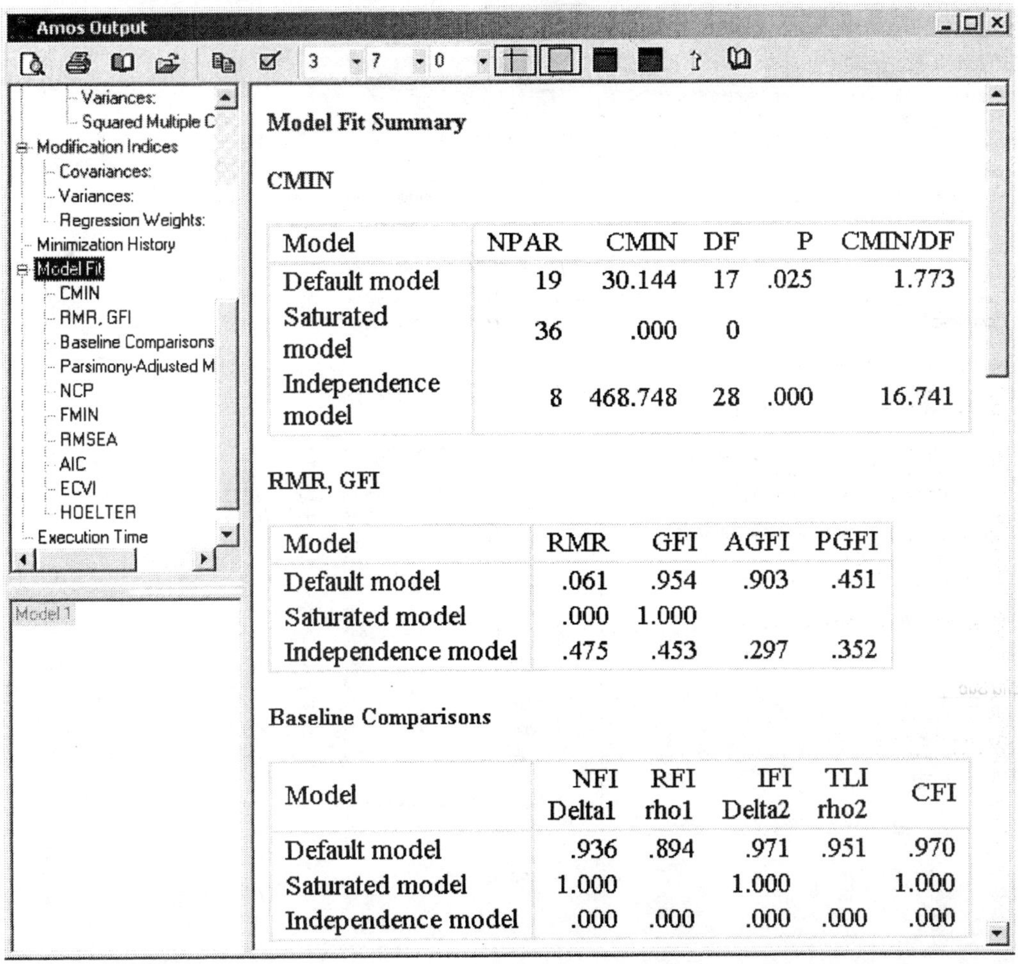

Amos Output

Variances:
Squared Multiple C
Modification Indices
Covariances:
Variances:
Regression Weights:
Minimization History
Model Fit
CMIN
RMR, GFI
Baseline Comparisons
Parsimony-Adjusted M
NCP
FMIN
RMSEA
AIC
ECVI
HOELTER
Execution Time

Model 1

Model Fit Summary

CMIN

Model	NPAR	CMIN	DF	P	CMIN/DF
Default model	19	30.144	17	.025	1.773
Saturated model	36	.000	0		
Independence model	8	468.748	28	.000	16.741

RMR, GFI

Model	RMR	GFI	AGFI	PGFI
Default model	.061	.954	.903	.451
Saturated model	.000	1.000		
Independence model	.475	.453	.297	.352

Baseline Comparisons

Model	NFI Delta1	RFI rho1	IFI Delta2	TLI rho2	CFI
Default model	.936	.894	.971	.951	.970
Saturated model	1.000		1.000		1.000
Independence model	.000	.000	.000	.000	.000

Maximum Likelihood Estimates and Standard Errors

The parameter estimates (non-standardized, estimated from the covariance matrix) are shown below. James Lattin, author of Analyzing Multivariate Data, comments, "The column 'C.R.' reports the critical ratio, given by the estimate divided by its standard error (S.E.). This is effectively a z-score, since the maximum likelihood routine provides asymptotic standard errors. These results suggest (and the standardized solution confirms) that 'interaction' has a greater impact on 'attitude' (in terms of explaining variance in the construct) than 'similarity.'"

Amos Output

Scalar Estimates (Group number 1 - Model 1)

Maximum Likelihood Estimates

Regression Weights: (Group number 1 - Model 1)

			Estimate	S.E.	C.R.	P	Label
attitude	<---	similarity	.177	.071	2.489	.013	
attitude	<---	interaction	1.042	.213	4.891	***	
Y1	<---	attitude	1.000				
Y2	<---	attitude	1.537	.183	8.385	***	
X1	<---	similarity	1.000				
X2	<---	similarity	.967	.128	7.530	***	
X3	<---	similarity	.986	.137	7.185	***	
X4	<---	interaction	1.000				
X5	<---	interaction	2.356	.380	6.198	***	
X6	<---	interaction	2.487	.384	6.472	***	

Standardized Regression Weights: (Group number 1 - Model 1)

			Estimate
attitude	<---	similarity	.254
attitude	<---	interaction	.647
Y1	<---	attitude	.833
Y2	<---	attitude	.756
X1	<---	similarity	.683
X2	<---	similarity	.825
X3	<---	similarity	.716
X4	<---	interaction	.552
X5	<---	interaction	.737
X6	<---	interaction	.857

Amos Output

Covariances: (Group number 1 - Model 1)

			Estimate	S.E.	C.R.	P	L
similarity	<-->	interaction	.191	.051	3.728	***	

Correlations: (Group number 1 - Model 1)

			Estimate
similarity	<-->	interaction	.513

Variances: (Group number 1 - Model 1)

	Estimate	S.E.	C.R.	P	Label
similarity	.858	.200	4.281	***	
interaction	.162	.048	3.379	***	
zeta	.146	.044	3.350	***	
eps1	.185	.044	4.192	***	
eps2	.741	.125	5.920	***	
eps3	.979	.144	6.799	***	
eps4	.377	.088	4.265	***	
eps5	.793	.124	6.370	***	
eps6	.368	.046	7.919	***	
eps7	.756	.117	6.444	***	
eps8	.360	.093	3.888	***	

Squared Multiple Correlations: (Group number 1 - Model 1)

	Estimate
attitude	.651
X6	.735
X5	.543
X4	.305
X3	.513
X2	.680
X1	.467
Y2	.572
Y1	.694

11 Analysis of Variance Using SPSS

Example 11.1 *Analysis of Variance of Finnish liquor data*

The Finnish liquor sales ANOVA tests whether mean rates of traffic accidents vary across 3 experimental treatment groups (municipalities with package liquor sales only, municipalities with both package and restaurant sales permitted, and a control group of municipalities where no liquor sales were allowed). Open Finland.sav as the working data file and rename variables col1, col2, col3 to Treat, Y and Z.

Open the **Analyze-General Linear Model-Univariate** menu path. Specify Y as the dependent variable (for accident rate) and Treat as the factor (for treatment type).

Click **OK** too run or **Paste** to display commands in the Syntax Editor.

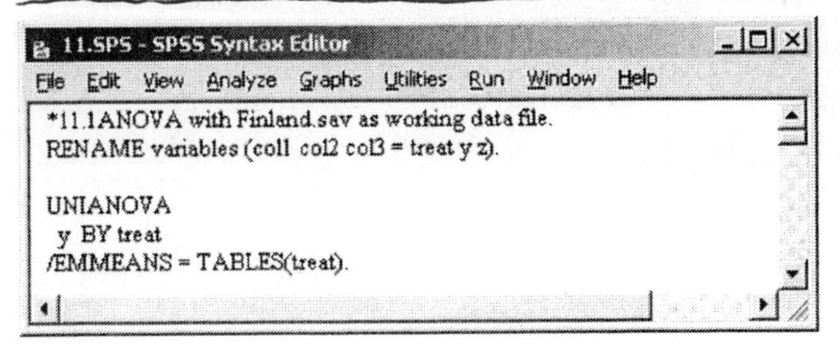

Output 11.1 *Results from ANOVA of Finnish liquor data*

The F value of 2.08 is not significant at the 0.05 level (sig = .181). However, treatment group means (groups 2 and 3) are slightly higher than the control group mean.

Tests of Between-Subjects Effects

Dependent Variable: Y

Source	Type III Sum of Squares	df	Mean Square	F	Sig.
Corrected Model	1696.167[a]	2	848.083	2.079	.181
Intercept	491670.083	1	491670.083	1205.484	.000
TREAT	1696.167	2	848.083	2.079	.181
Error	3670.750	9	407.861		
Total	497037.000	12			
Corrected Total	5366.917	11			

a. R Squared = .316 (Adjusted R Squared = .164)

TREAT

Dependent Variable: Y

TREAT	Mean	Std. Error	95% Confidence Interval	
			Lower Bound	Upper Bound
1	186.250	10.098	163.407	209.093
2	206.500	10.098	183.657	229.343
3	214.500	10.098	191.657	237.343

Example 11.2 *Analysis of Covariance of Finnish Liquor Data.*

This analysis of covariance statistically controls for differences across treatment groups by including covariate Z, number of accidents in the year prior to the test period.

Click the **Options** button to obtain Estimated Marginal Means output for variable TREAT.

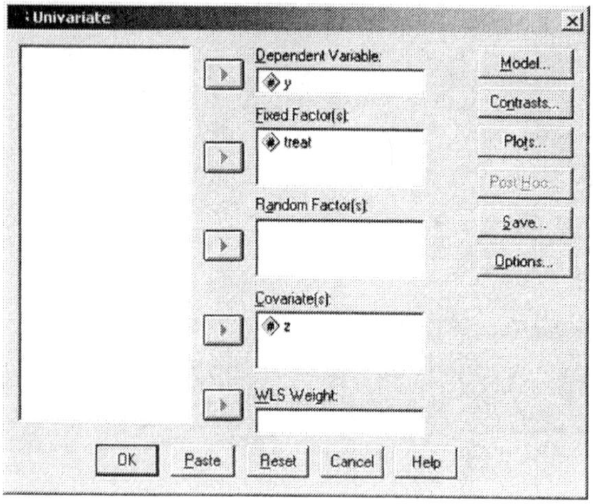

Univariate

Dependent Variable: y
Fixed Factor(s): treat
Random Factor(s):
Covariate(s): z
WLS Weight:
Model... Contrasts... Plots... Post Hoc... Save... Options...
OK Paste Reset Cancel Help

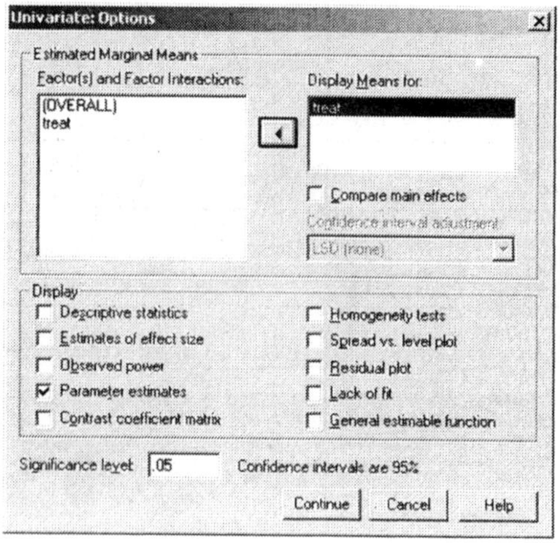

Univariate: Options

Estimated Marginal Means
Factor(s) and Factor Interactions: (OVERALL) treat
Display Means for: treat
Compare main effects
Confidence interval adjustment: LSD (none)
Display
Descriptive statistics
Estimates of effect size
Observed power
Parameter estimates
Contrast coefficient matrix
Homogeneity tests
Spread vs. level plot
Residual plot
Lack of fit
General estimable function
Significance level: .05 Confidence intervals are 95%
Continue Cancel Help

Completed ANCOVA syntax

11 - SPSS Syntax Editor

File Edit View Analyze Graphs Utilities Run Window Help

```
*ANCOVA Finnish Liquor Policy Experiment
UNIANOVA
  y BY treat WITH z
  /EMMEANS = TABLES(treat) WITH(z=MEAN)
  /PRINT = PARAMETER.
```

SPSS Processor i

Output 11.2 *Results from ANCOVA of Finnish liquor data*

Note the F score for treatment is now 12.8 and significant to the 0.01level.

Tests of Between-Subjects Effects

Dependent Variable: Y

Source	Type III Sum of Squares	df	Mean Square	F	Sig.
Corrected Model	4618.238[a]	3	1539.413	16.449	.001
Intercept	179.386	1	179.386	1.917	.204
Z	2922.071	1	2922.071	31.224	.001
TREAT	2401.757	2	1200.878	12.832	.003
Error	748.679	8	93.585		
Total	497037.000	12			
Corrected Total	5366.917	11			

a. R Squared = .861 (Adjusted R Squared = .808)

The adjusted mean accident rate of 223.6 for municipalities licensing both package stores and restaurants (TREAT = 3) suggests this condition increases accident rates the most.

TREAT

Dependent Variable: Y

TREAT	Mean	Std. Error	95% Confidence Interval	
			Lower Bound	Upper Bound
1	191.172[a]	4.917	179.835	202.510
2	192.462[a]	5.450	179.893	205.031
3	223.616[a]	5.105	211.844	235.387

a. Evaluated at covariates appeared in the model: Z = 222.25.

Example 11.3 *ANCOVA on Newfood test data.*

Open Newfood.sav as the working data file. Compute a new dependent variable, Y, to represent total sales over 6 months.

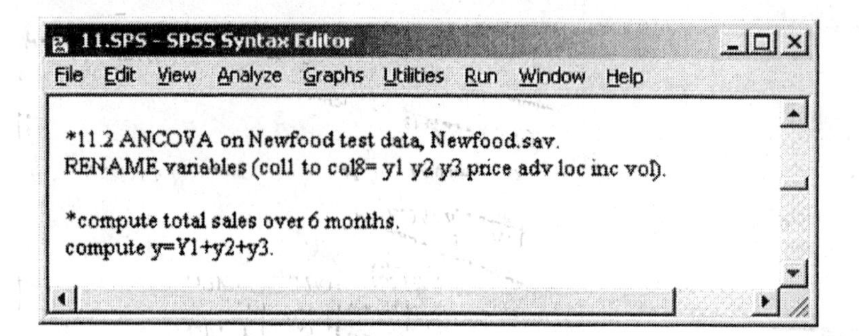

Example 11.2 demonstrates the SPSS GLM command. SPSS GLM is a menu-driven general procedure for analysis of variance, covariance and regression. Individual or multiple dependent variables can be analyzed by one or more factors or independent variables. The GLM command provides estimates of parameters, tests of effects of individual factors, tests of interaction between factors and tests of covariate effects. The basic format of GLM lists dependent variable, factors (if any) and covariates (if any):

GLM By *factor1 factor2* WITH *convariate1 covariate2*

Select Y as dependent variable, Adv (advertising) and Price as factors, and vol (volume) as covariate. Click the **Model** button to define Model elements.

Click the **Options** button to specify output; including estimated marginal means for adv, price and the adv-price interaction.

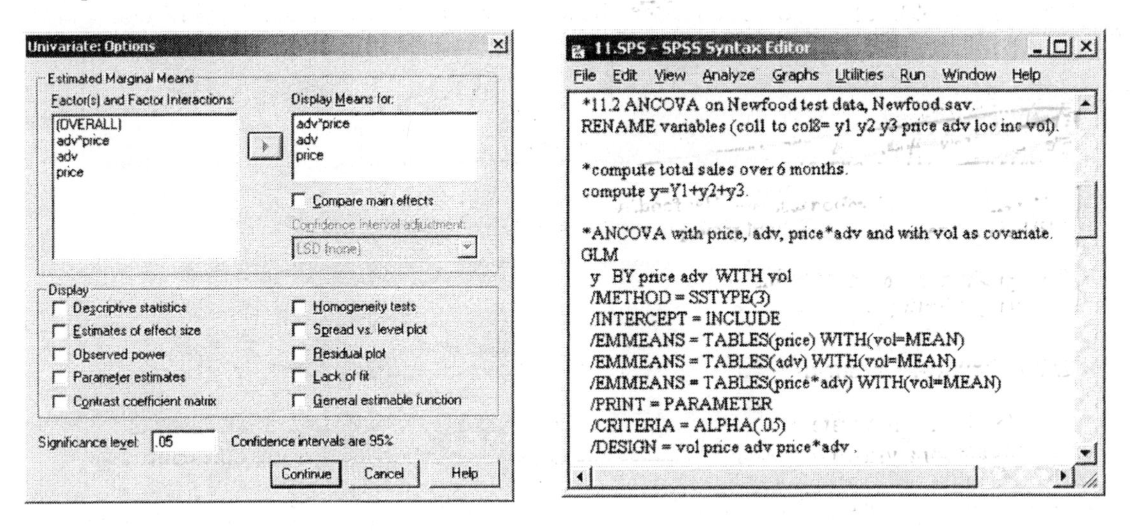

Output 11.3 *Results from ANCOVA of Newfood data*

Tests of Between-Subjects Effects

Dependent Variable: Y

Source	Type III Sum of Squares	df	Mean Square	F	Sig.
Corrected Model	866760.344[a]	6	144460.057	14.188	.000
Intercept	8742.066	1	8742.066	.859	.367
ADV * PRICE	45304.159	2	22652.079	2.225	.139
ADV	113931.737	1	113931.737	11.189	.004
PRICE	393082.793	2	196541.396	19.302	.000
VOL	192465.011	1	192465.011	18.902	.000
Error	173097.489	17	10182.205		
Total	8876266.000	24			
Corrected Total	1039857.833	23			

a. R Squared = .834 (Adjusted R Squared = .775)

The ADV * Price table shows mean sales adjusted for store size.

1. ADV * PRICE

Dependent Variable: Y

ADV	PRICE	Mean	Std. Error	95% Confidence Interval	
				Lower Bound	Upper Bound
1	1	581.973[a]	57.104	461.494	702.451
	2	448.470[a]	58.361	325.340	571.601
	3	365.233[a]	53.049	253.309	477.158
2	1	909.512[a]	51.879	800.057	1018.968
	2	642.031[a]	58.799	517.975	766.086
	3	481.281[a]	58.799	357.225	605.336

a. Evaluated at covariates appeared in the model: VOL = 32.33.

Example 11.4 *Multiple analysis of Variance (MANOVA), Testing advertising strategy: ad copy effects on likeability and purchase intent*

Example 11.3 demonstrates the SPSS GLM command used to produce a Multiple Analysis of Variance. See Example 11.2 for introduction to GLM.

The dependent variables are Like (preference for product) and Intent (intention to buy). Treat is the independent variable, or "Fixed Factor" (1 = hard sell ad and 2= humorous ad). Click the **Model** button to specify the type of model.

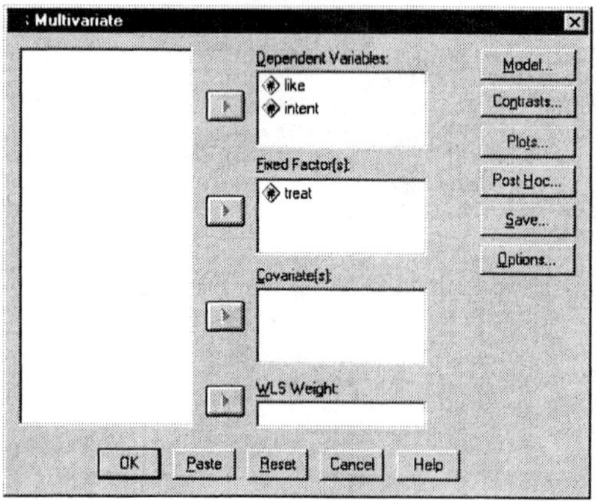

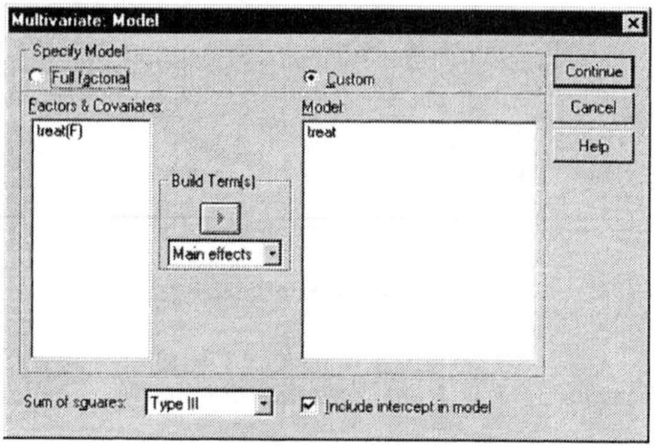

Click the **Contrasts** button to compare a Contrast Estimate (a linear combination of cell means) to a hypothesized value (default zero). Specify estimated marginal means output for variable Treat.

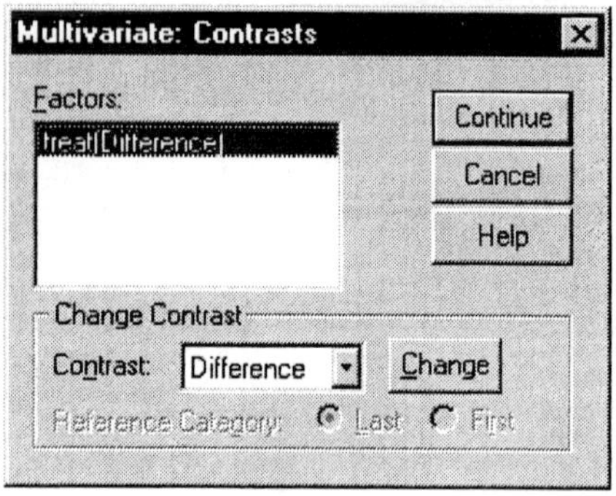

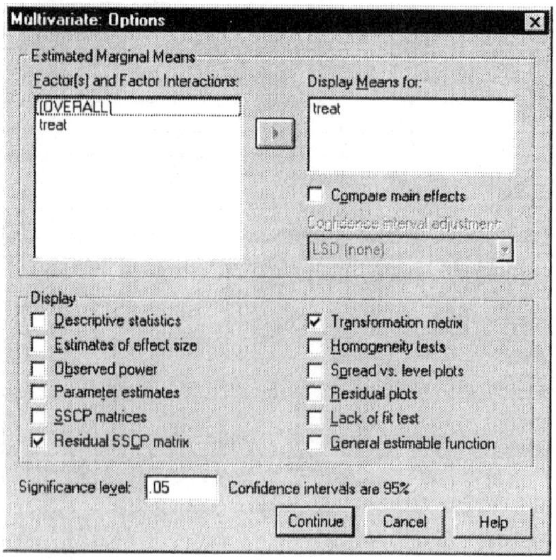

Syntax for SPSS GLM and alternate SPSS MANOVA command syntax

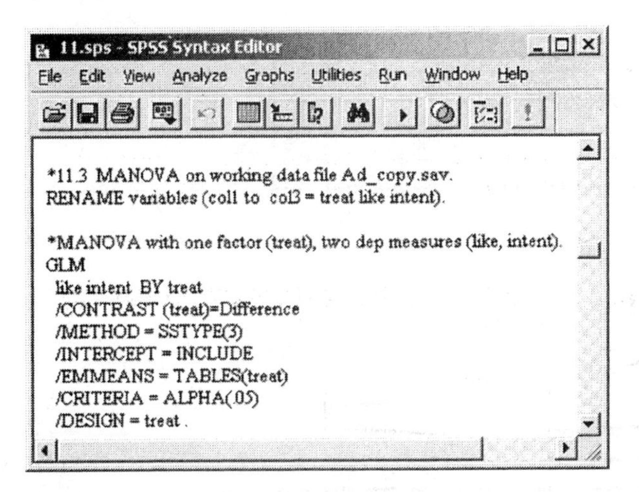

An alternate procedure for Multiple Analysis of Variance is the SPSS MANOVA command. The SPSS MANOVA command is not available through menus and can be created only by text entry in a syntax file. MANOVA is not demonstrated in this manual. See SPSS Help for a MANOVA syntax diagram.

Output 11.4 *Results from GLM univariate ANOVAs of ad copy test data*

Tests of Between-Subjects Effects

Source	Dependent Variable	Type III Sum of Squares	df	Mean Square	F	Sig.
Corrected Model	INTENT	2.817[a]	1	2.817	2.796	.100
	LIKE	3.750[b]	1	3.750	2.846	.097
Intercept	INTENT	1353.750	1	1353.750	1343.711	.000
	LIKE	1372.817	1	1372.817	1041.736	.000
TREAT	INTENT	2.817	1	2.817	2.796	.100
	LIKE	3.750	1	3.750	2.846	.097
Error	INTENT	58.433	58	1.007		
	LIKE	76.433	58	1.318		
Total	INTENT	1415.000	60			
	LIKE	1453.000	60			
Corrected Total	INTENT	61.250	59			
	LIKE	80.183	59			

a. R Squared = .046 (Adjusted R Squared = .030)

b. R Squared = .047 (Adjusted R Squared = .030)

Contrast Results (K Matrix)

TREAT Difference Contrast			Dependent Variable INTENT	Dependent Variable LIKE
Level 2 vs. Level 1	Contrast Estimate		-.433	.500
	Hypothesized Value		0	0
	Difference (Estimate - Hypothesized)		-.433	.500
	Std. Error		.259	.296
	Sig.		.100	.097
	95% Confidence Interval for Difference	Lower Bound	-.952	-9.331E-02
		Upper Bound	8.544E-02	1.093

Multivariate Test Results

	Value	F	Hypothesis df	Error df	Sig.
Pillai's trace	.257	9.838[a]	2.000	57.000	.000
Wilks' lambda	.743	9.838[a]	2.000	57.000	.000
Hotelling's trace	.345	9.838[a]	2.000	57.000	.000
Roy's largest root	.345	9.838[a]	2.000	57.000	.000

a. Exact statistic

Univariate Test Results

Source	Dependent Variable	Sum of Squares	df	Mean Square	F	Sig.
Contrast	INTENT	2.817	1	2.817	2.796	.100
	LIKE	3.750	1	3.750	2.846	.097
Error	INTENT	58.433	58	1.007		
	LIKE	76.433	58	1.318		

Treat 1 (hard sell) has higher intent to buy but lower likeability than Treat 2 (humorous ad).

TREAT

				95% Confidence Interval	
Dependent Variable	TREAT	Mean	Std. Error	Lower Bound	Upper Bound
INTENT	1	4.967	.183	4.600	5.333
	2	4.533	.183	4.167	4.900
LIKE	1	4.533	.210	4.114	4.953
	2	5.033	.210	4.614	5.453

Example 11.5 *Repeated measures designs with MANOVA*

Use Newfood.sav as the working data file for this example. Sales volume is broken down into three sales variables: Y_1, Y_2, and Y_3, representing sales at 2, 4 and 6 months. Following the menu path **General Linear Model-Repeated Measures,** define the three-level within-subject factor 'Time'. Specify 3 levels and then click the **Define** button.

In the **Repeated Measures** dialog box match variables Y_1, Y_2, and Y_3 to the three levels of the Time factor. Define Adv and Price as between-subjects factors. Define Vol (volume) as a covariate.

GLM repeated measures syntax

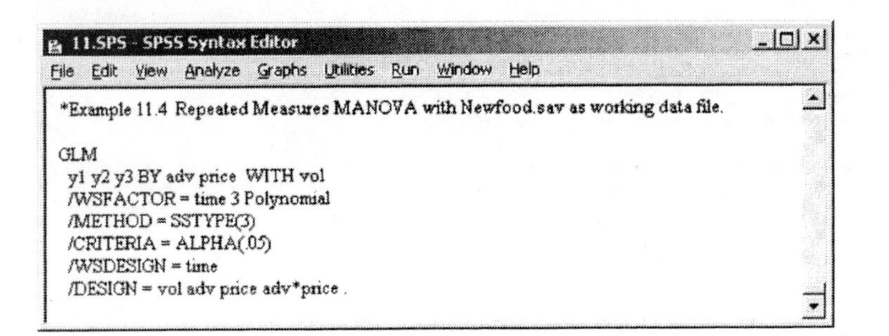

11.SPS - SPSS Syntax Editor

```
*Example 11.4 Repeated Measures MANOVA with Newfood.sav as working data file.

GLM
  y1 y2 y3 BY adv price WITH vol
  /WSFACTOR = time 3 Polynomial
  /METHOD = SSTYPE(3)
  /CRITERIA = ALPHA(.05)
  /WSDESIGN = time
  /DESIGN = vol adv price adv*price .
```

Output 11.5 *Multivariate tests of model effects, test of sphericity, Within-subject effects and Between-subjects effects*

Time and the interaction effect between time and advertising have statistically significant effects on sales pattern (the significant interaction effect of time and volume reflects the higher volume of early sales). Wilks Lambda values close to one indicate no significant difference between group means. For example, the time-price and time-price-advertising interactions have Wilks values of .810 and .700 respectively and are not significant by any measure.

Multivariate Tests [c]

Effect		Value	F	Hypothesis df	Error df	Sig.
TIME	Pillai's Trace	.477	7.282[a]	2.000	16.000	.006
	Wilks' Lambda	.523	7.282[a]	2.000	16.000	.006
	Hotelling's Trace	.910	7.282[a]	2.000	16.000	.006
	Roy's Largest Root	.910	7.282[a]	2.000	16.000	.006
TIME * VOL	Pillai's Trace	.658	15.422[a]	2.000	16.000	.000
	Wilks' Lambda	.342	15.422[a]	2.000	16.000	.000
	Hotelling's Trace	1.928	15.422[a]	2.000	16.000	.000
	Roy's Largest Root	1.928	15.422[a]	2.000	16.000	.000
TIME * ADV	Pillai's Trace	.711	19.655[a]	2.000	16.000	.000
	Wilks' Lambda	.289	19.655[a]	2.000	16.000	.000
	Hotelling's Trace	2.457	19.655[a]	2.000	16.000	.000
	Roy's Largest Root	2.457	19.655[a]	2.000	16.000	.000
TIME * PRICE	Pillai's Trace	.190	.890	4.000	34.000	.481
	Wilks' Lambda	.810	.886[a]	4.000	32.000	.483
	Hotelling's Trace	.234	.877	4.000	30.000	.489
	Roy's Largest Root	.234	1.987[b]	2.000	17.000	.168
TIME * ADV * PRICE	Pillai's Trace	.311	1.567	4.000	34.000	.205
	Wilks' Lambda	.700	1.560[a]	4.000	32.000	.209
	Hotelling's Trace	.411	1.542	4.000	30.000	.215
	Roy's Largest Root	.366	3.108[b]	2.000	17.000	.071

a. Exact statistic

b. The statistic is an upper bound on F that yields a lower bound on the significance level.

c.

Design: Intercept+VOL+ADV+PRICE+ADV * PRICE
Within Subjects Design: TIME

The Mauchly's test value of .770 means the null hypothesis for sphericity is not rejected in this case.

Mauchly's Test of Sphericity [b]

Measure: MEASURE_1

					Epsilon[a]		
Within Subjects Effect	Mauchly's W	Approx. Chi-Square	df	Sig.	Greenhouse-Geisser	Huynh-Feldt	Lower-bound
TIME	.968	.523	2	.770	.969	1.000	.500

Tests the null hypothesis that the error covariance matrix of the orthonormalized transformed dependent variables is proportional to an identity matrix.

a. May be used to adjust the degrees of freedom for the averaged tests of significance. Corrected tests are displayed in the Tests of Within-Subjects Effects table.

b.

Design: Intercept+VOL+ADV+PRICE+ADV * PRICE
Within Subjects Design: TIME

The Tests of Within-Subjects Table displays univariate tests and interaction terms. Refer to the Sphericity Assumed row where the Mauchly's test null is not rejected. Wherever sphericity is assumed assumptions about the variance-covariance matrix can based on the original degrees of freedom.

Tests of Within-Subjects Effects

Measure: MEASURE_1

Source		Type III Sum of Squares	df	Mean Square	F	Sig.
TIME	Sphericity Assumed	5643.860	2	2821.930	8.602	.001
	Greenhouse-Geisser	5643.860	1.938	2912.661	8.602	.001
	Huynh-Feldt	5643.860	2.000	2821.930	8.602	.001
	Lower-bound	5643.860	1.000	5643.860	8.602	.009
TIME * VOL	Sphericity Assumed	11418.145	2	5709.073	17.402	.000
	Greenhouse-Geisser	11418.145	1.938	5892.631	17.402	.000
	Huynh-Feldt	11418.145	2.000	5709.073	17.402	.000
	Lower-bound	11418.145	1.000	11418.145	17.402	.001
TIME * ADV	Sphericity Assumed	12690.533	2	6345.266	19.342	.000
	Greenhouse-Geisser	12690.533	1.938	6549.279	19.342	.000
	Huynh-Feldt	12690.533	2.000	6345.266	19.342	.000
	Lower-bound	12690.533	1.000	12690.533	19.342	.000
TIME * PRICE	Sphericity Assumed	1499.863	4	374.966	1.143	.353
	Greenhouse-Geisser	1499.863	3.875	387.022	1.143	.353
	Huynh-Feldt	1499.863	4.000	374.966	1.143	.353
	Lower-bound	1499.863	2.000	749.932	1.143	.342
TIME * ADV * PRICE	Sphericity Assumed	2217.045	4	554.261	1.689	.175
	Greenhouse-Geisser	2217.045	3.875	572.082	1.689	.177
	Huynh-Feldt	2217.045	4.000	554.261	1.689	.175
	Lower-bound	2217.045	2.000	1108.522	1.689	.214
Error(TIME)	Sphericity Assumed	11154.188	34	328.064		
	Greenhouse-Geisser	11154.188	32.941	338.612		
	Huynh-Feldt	11154.188	34.000	328.064		
	Lower-bound	11154.188	17.000	656.129		

The table on Tests of Between-Subjects Effects indicates that there are significant differences in sales volume, advertising and price at time 1, time 2 and time 3.

Tests of Between-Subjects Effects

Measure: MEASURE_1
Transformed Variable: Average

Source	Type III Sum of Squares	df	Mean Square	F	Sig.
Intercept	2914.022	1	2914.022	.859	.367
VOL	64155.004	1	64155.004	18.902	.000
ADV	37977.246	1	37977.246	11.189	.004
PRICE	131027.598	2	65513.799	19.302	.000
ADV * PRICE	15101.386	2	7550.693	2.225	.139
Error	57699.163	17	3394.068		

12 Discriminant Analysis Using SPSS

12.1 Two-group discriminant analysis
12.2 Computing the discriminant function
12.3 Analysis of holdout sample
12.4 Three-group discriminant analysis

This chapter demonstrates analyses using the SPSS DISCRIMINANT command.

Example 12.1 *Two-group discriminant analysis -- Books by Mail data*

Open Books_1.sav as the working data file. Rename variables col1-col4 to Custid (Customer ID), Recent (recency of purchase), Artbooks (number of art books purchased) and the grouping variable Buy (purchased=1, Did not Purchase=0). Run the FREQUENCIES command on variable Buy, where 0 means 'did not purchase' and 1 means 'purchased. 83 have purchased a book and 917 have not.

BUY

		Frequency	Percent	Valid Percent	Cumulative Percent
Valid	0	917	91.7	91.7	91.7
	1	83	8.3	8.3	100.0
	Total	1000	100.0	100.0	

In the **Analyze** menu select **Classify-Discriminant** to open the **Discriminant** Analysis dialog box. Define Buy as the grouping variable. Specify the classification range as two groups (0 and 1), by clicking the **Define Range** button.

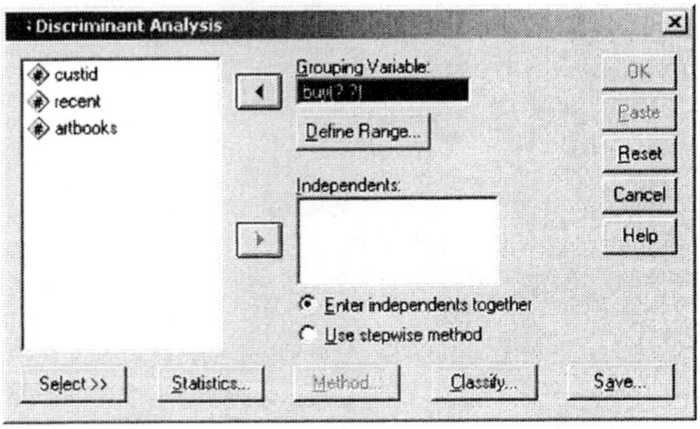

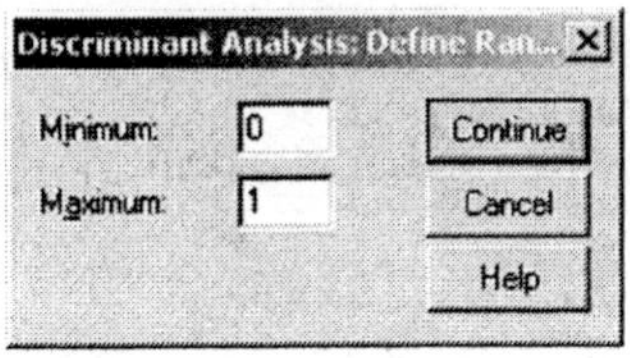

Select Recent and Artbooks as independent variables. In the **Statistics** dialog box select checkboxes for descriptive statistics and for unstandardized coefficients output. Unstandardized coefficients can be used to compute the discriminant function (see Example 12.2).

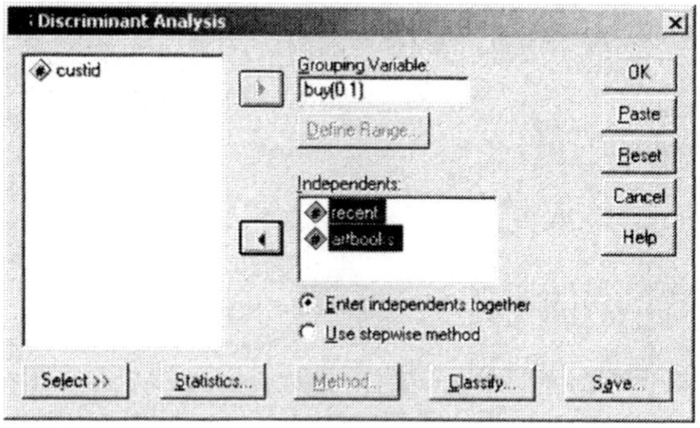

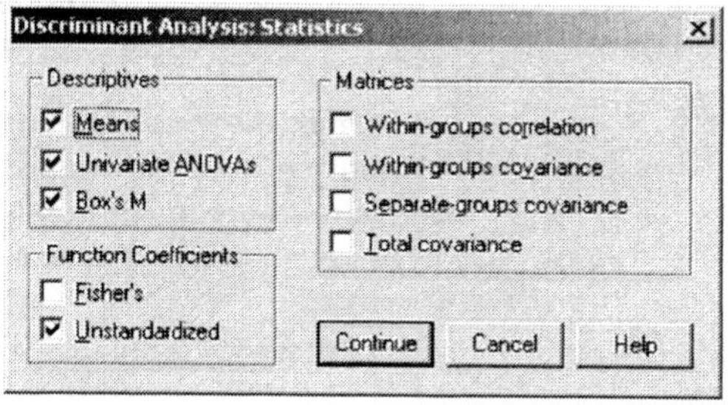

In the **Classification** dialog box define classification rules and display of results. For this analysis check the button to compute prior probabilities from group sizes. Thus non-purchasers will have approximately 92% prior probability, because the group contains 917 of 1000 cases. Check boxes to display casewise results for the first five cases and to display a Summary Table.

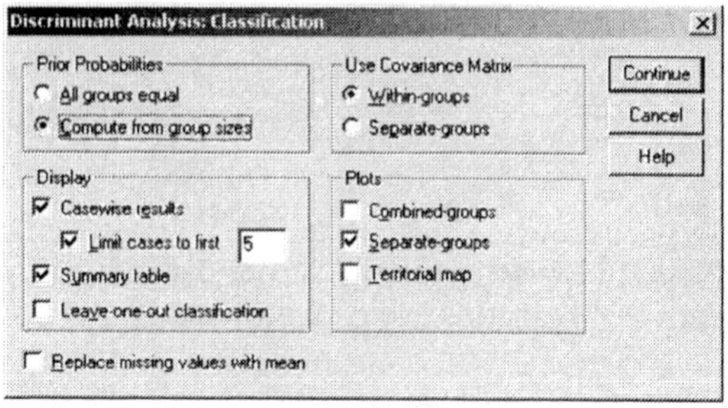

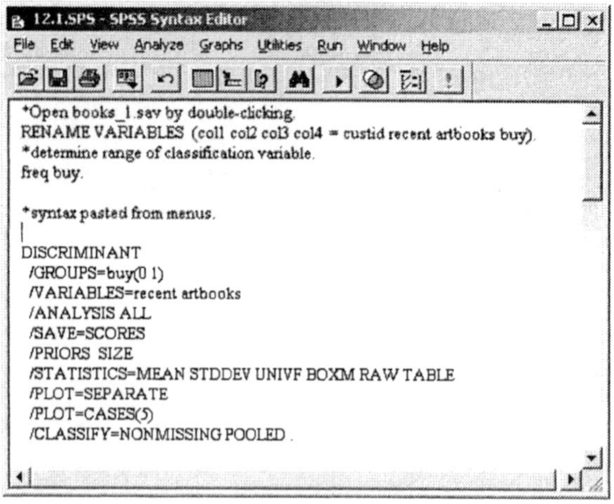

```
*Open books_1.sav by double-clicking.
RENAME VARIABLES (col1 col2 col3 col4 = custid recent artbooks buy).
*determine range of classification variable.
freq buy.

*syntax pasted from menus.

DISCRIMINANT
  /GROUPS=buy(0 1)
  /VARIABLES=recent artbooks
  /ANALYSIS ALL
  /SAVE=SCORES
  /PRIORS SIZE
  /STATISTICS=MEAN STDDEV UNIVF BOXM RAW TABLE
  /PLOT=SEPARATE
  /PLOT=CASES(5)
  /CLASSIFY=NONMISSING POOLED .
```

SPSS DISCRIMINANT has two required subcommands: the /GROUPS keyword and specification, and the /VARIABLES keyword and specification.

OUTPUT 12.1.A *Summary Statistics*

The Group Statistics table displays univariate statistics, including means, standard deviations and valid cases. Discriminant analysis assumes equal variances, so the standard deviations should show low or moderate variation across groups. In this example the standard deviation for Recent varies between 5.95 for purchasers to 8.11 for non-purchasers. The Box's M statistic with significance less than .05 suggests that there is a significant difference between group covariance matrices.

Group Statistics

BUY		Mean	Std. Deviation	Valid N (listwise) Unweighted	Weighted
0	RECENT	12.73	8.11	917	917.000
	ARTBOOKS	.33	.61	917	917.000
1	RECENT	9.41	5.95	83	83.000
	ARTBOOKS	1.00	1.06	83	83.000
Total	RECENT	12.46	8.00	1000	1000.000
	ARTBOOKS	.39	.68	1000	1000.000

Test Results

Box's M		77.809
F	Approx.	25.706
	df1	3
	df2	243266.705
	Sig.	.000

Tests null hypothesis of equal population covariance matrices.

The canonical correlation in the Eigenvalues Table is an overall measure of association between each group and discriminant scores.

Eigenvalues

Function	Eigenvalue	% of Variance	Cumulative %	Canonical Correlation
1	.094[a]	100.0	100.0	.293

a. First 1 canonical discriminant functions were used in the analysis.

The Wilks' Lambda Table tests the null hypothesis of equivalent group means. The Lambda value is the proportion of total variance of scores not explained by group mean differences. The .914 value

Wilks' Lambda

Test of Function(s)	Wilks' Lambda	Chi-square	df	Sig.
1	.914	89.611	2	.000

(nearly the maximum of 1) and low significance level suggest the group means are different.

Unstandardized canonical discriminant function coefficients are the terms for computing case-by-case canonical scores to predict group membership. Standardized coefficients values indicate the relative predictive importance of independent variables.

Canonical Discriminant Function Coefficients

	Function
	1
RECENT	-.051
ARTBOOKS	1.412
(Constant)	.086

Unstandardized coefficients

Standardized Canonical Discriminant Function Coefficients

	Function
	1
RECENT	-.405
ARTBOOKS	.927

The structure matrix lists the within-group correlations of each independent variable, indicates their usefulness in an analysis with multiple discriminant functions. The structure matrix is uninteresting in this single-function analysis. The Functions at Group Centroids Table displays within-group mean canonical scores. The mean canonical score for those who bought books is 1.018.

Structure Matrix

	Function
	1
ARTBOOKS	.914
RECENT	-.376

Pooled within-groups correlations between discriminating variables and standardized canonical discriminant functions
Variables ordered by absolute size of correlation within function.

Functions at Group Centroids

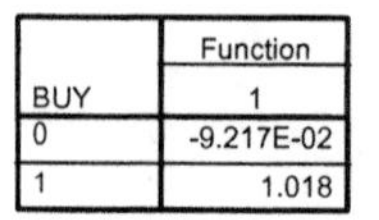

BUY	Function 1
0	-9.217E-02
1	1.018

Unstandardized canonical discriminant functions evaluated at group means

Histograms of Group Distributions (Note that vertical scales are different)

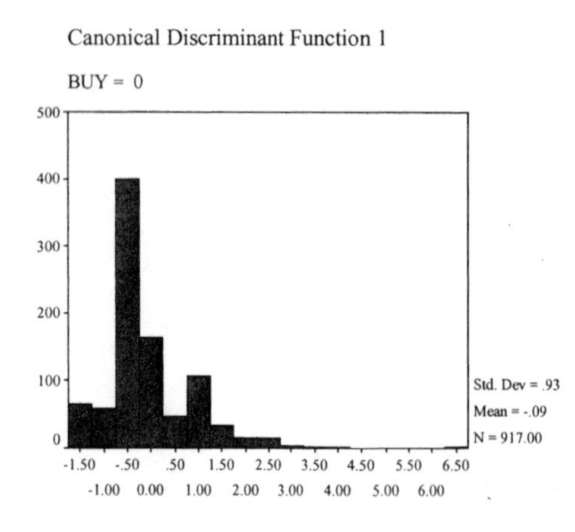

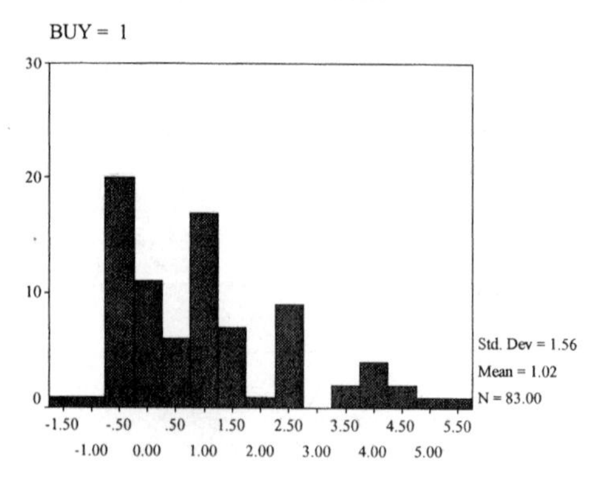

OUTPUT 12.1.B *Classification Statistics*

The Prior Probabilities for Groups table lists prior probabilities and their weighting effects. The Specified Prior and Effective Prior cells are empty below because prior probabilities in this analysis are proportional to group size.

Prior Probabilities for Groups

BUY	Prior	Specified Prior	Effective Prior	Cases Used in Analysis	
				Unweighted	Weighted
0	.917			917	917.000
1	.083			83	83.000
Total	1.000			1000	1000.000

The Casewise Statistics Table, specified by the /PLOT=CASES (5) subcommand, displays predicted and actual classification performance for the first five cases of the dataset. Four Non-buyers are correctly classified and one Buyer (Group 1) is misclassified. Posterior probabilities are recorded in the column labeled P(G=g) | D=d). Canonical discriminant scores are recorded in the far right column.

Casewise Statistics

	Case Number	Actual Group	Highest Group					Second Highest Group			Discriminant Scores
			Predicted Group	P(D>d | G=g)		P(G=g | D=d)	Squared Mahalanobis Distance to Centroid	Group	P(G=g | D=d)	Squared Mahalanobis Distance to Centroid	Function 1
				p	df						
Original	1	0	0	.296	1	.985	1.094	1	.015	4.649	-1.138
	2	0	0	.524	1	.977	.407	1	.023	3.057	-.730
	3	0	0	.557	1	.975	.345	1	.025	2.881	-.679
	4	0	0	.345	1	.983	.891	1	.017	4.220	-1.036
	5	1	0**	.557	1	.975	.345	1	.025	2.881	-.679

**. Misclassified case

The Classification Results Table summarizes the effectiveness of this discriminant function as a classification rule. Count and percentage of correct and incorrect case classifications are displayed. This analysis classified 92% of cases correctly. Ten Non-Buyers were incorrectly classified as Buyers, 72 Buyers were incorrectly classified as Non-Buyers.

Classification Results [a]

		BUY	Predicted Group Membership		Total
			0	1	
Original	Count	0	907	10	917
		1	72	11	83
	%	0	98.9	1.1	100.0
		1	86.7	13.3	100.0

a. 91.8% of original grouped cases correctly classified.

Example 12.2 *Computing the discriminant function -- Books by Mail data*

Canonical scores can be calculated and saved to the working data file in two ways. First, canonical scores can be calculated automatically by including the /SAVE SCORES subcommand within the SPSS DISCRIMINANT command. From the **Analyze** menu select **Classify-Discriminant**, and then click the **Save** button.

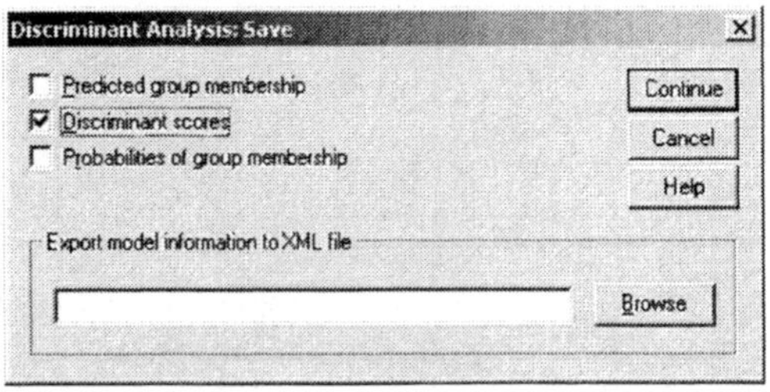

Discriminant Analysis: Save

- [ ] Predicted group membership
- [x] Discriminant scores
- [ ] Probabilities of group membership

Continue | Cancel | Help

Export model information to XML file

Browse

Second, canonical scores can be calculated within the SPSS COMPUTE command by defining an equation based on unstandardized canonical discriminant coefficients. These are displayed in Output 12.1.A, the Canonical Discriminant Function Coefficients table. The equation to calculate canonical scores: Score = (-.051* recent) + (1.412*artbooks) + .086

The following syntax demonstrates both approaches to calculating canonical scores. Each adds identical canonical discriminant score variables to the working dataset. The variable created by the /SAVE=SCORES subcommand is automatically named 'dis1_1.' The variable created by the SPSS COMPUTE command is a user-defined name, in this example 'Score.'

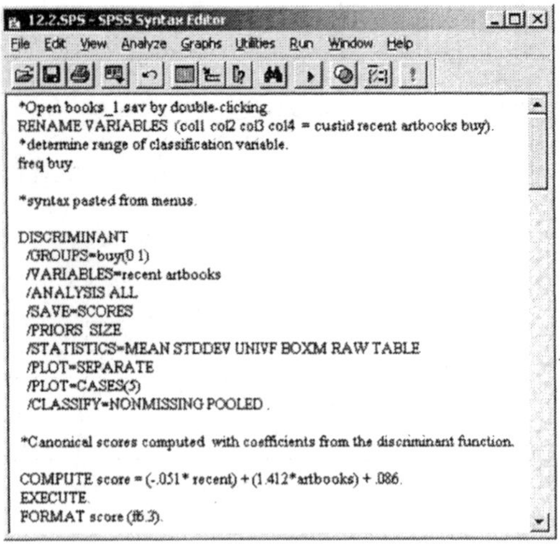

12.2.SPS - SPSS Syntax Editor

```
*Open books_1.sav by double-clicking
RENAME VARIABLES (col1 col2 col3 col4 = custid recent artbooks buy).
*determine range of classification variable.
freq buy.

*syntax pasted from menus.

DISCRIMINANT
 /GROUPS=buy(0 1)
 /VARIABLES=recent artbooks
 /ANALYSIS ALL
 /SAVE=SCORES
 /PRIORS SIZE
 /STATISTICS=MEAN STDDEV UNIVF BOXM RAW TABLE
 /PLOT=SEPARATE
 /PLOT=CASES(5)
 /CLASSIFY=NONMISSING POOLED .

*Canonical scores computed with coefficients from the discriminant function.

COMPUTE score = (-.051* recent) + (1.412*artbooks) + .086.
EXECUTE.
FORMAT score (f6.3).
```

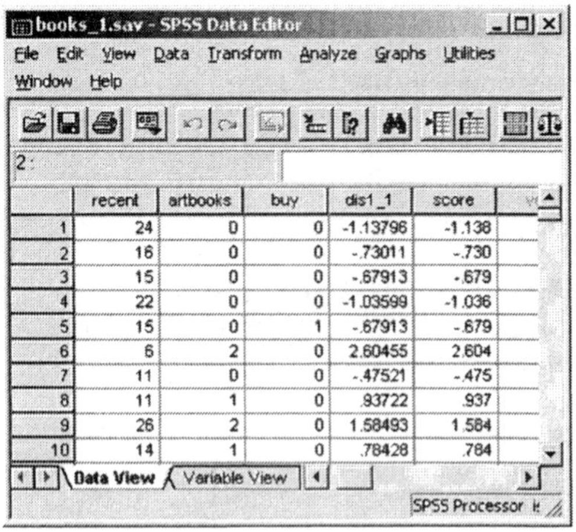

books_1.sav - SPSS Data Editor

	recent	artbooks	buy	dis1_1	score
1	24	0	0	-1.13796	-1.138
2	16	0	0	-.73011	-.730
3	15	0	0	-.67913	-.679
4	22	0	0	-1.03599	-1.036
5	15	0	1	-.67913	-.679
6	6	2	0	2.60455	2.604
7	11	0	0	-.47521	-.475
8	11	1	0	.93722	.937
9	26	2	0	1.58493	1.584
10	14	1	0	.78428	.784

Example 12.3 *Analysis of holdout sample with priors from validation sample -- Books by Mail data*

This example demonstrates validation of Discriminant analysis by comparison to a holdout sample, the Books_2.sav data file. First assess classification accuracy by running a discriminant analysis on Books2.sav using priors obtained from analysis of the calibration sample Books_1.sav (Example 12.1). The Classification Results Table demonstrates that calibration sample priors have similar predictive results for both calibration and holdout samples.

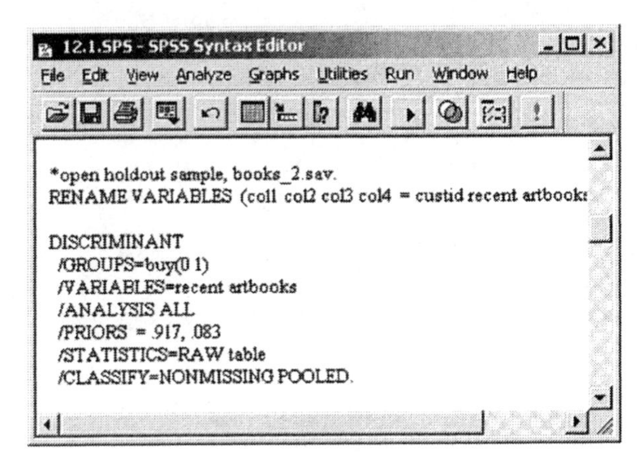

Classification Results [a]

			Predicted Group Membership		
		BUY	0	1	Total
Original	Count	0	894	25	919
		1	63	18	81
	%	0	97.3	2.7	100.0
		1	77.8	22.2	100.0

a. 91.2% of original grouped cases correctly classified.

Second, run an analysis on Books2.sav in which priors reflect asymmetric costs of misclassification. Where asymmetric costs of misclassification are the basis for defining priors the rate of correct classification declines to 82%. However, the number of correctly classified customers quadruples, from 11 correctly identified in the original analysis to 44 correctly identified.

12.1.SPS - SPSS Syntax Editor

```
* Change books_1.sav priors to reflect the asymmetric costs of misclassification.
* Expected cost of non-purchase: $1 x 0.917 = 0.917 .
* Expected profit from purchase: $6 x 0.083 = 0.498.
  * Priors are .917/(.917+.498)=.648 and 1-.648=.352.

*rerun boosk2_sav.

DISCRIMINANT
 /GROUPS=buy(0 1)
 /VARIABLES=recent artbooks
 /ANALYSIS ALL
 /PRIORS = .648, .352
 /STATISTICS= table
 /CLASSIFY=NONMISSING POOLED
```

Classification Results [a]

			Predicted Group Membership		
		BUY	0	1	Total
Original	Count	0	780	139	919
		1	37	44	81
	%	0	84.9	15.1	100.0
		1	45.7	54.3	100.0

a. 82.4% of original grouped cases correctly classified.

Example 12.4 *Three-group analysis -- Real Estate listings*

Rename variables and convert data type to numeric

Open Real_estate.sav as the working dataset and rename variables Col1 through Col4 to District, Price, Bedrooms and Lot_size. Transform the grouping variable, District from string (character) format to the numeric format required by the DISCRIMINANT command. Within the SPSS Data Editor **Variable View** click on the relevant cell in the **Type** column to define the District variable type as Numeric. Select the **Numeric** radio button in the **Variable Type** dialog box, then click OK.

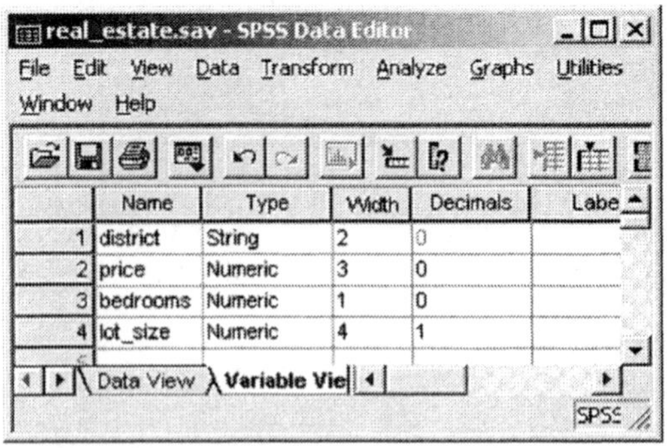

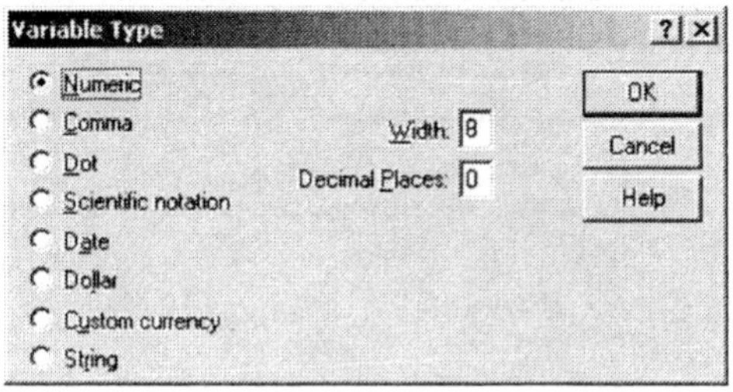

	Name	Type	Width	Decimals
1	district	String	2	0
2	price	Numeric	3	0
3	bedrooms	Numeric	1	0
4	lot_size	Numeric	4	1

	Name	Type	Width	Decimals
1	district	Numeric	2	0
2	price	Numeric	3	0
3	bedrooms	Numeric	1	0
4	lot_size	Numeric	4	1

3-group discriminant analysis syntax is similar to two-group, except that additional diagnostics are run to compare the several discriminant functions. Diagnostics for assessing the discriminant functions can be selected from the **Statistics** dialog box.

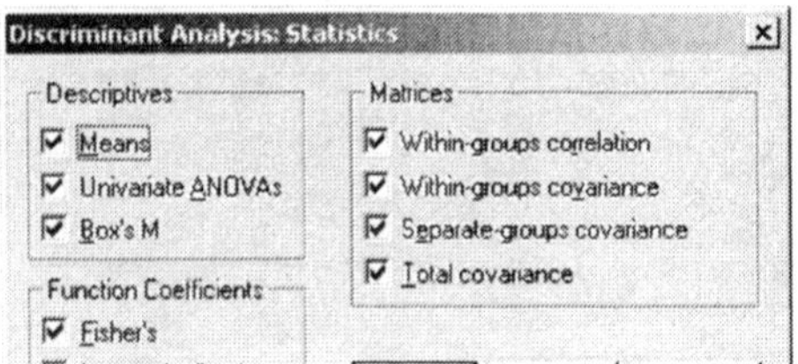

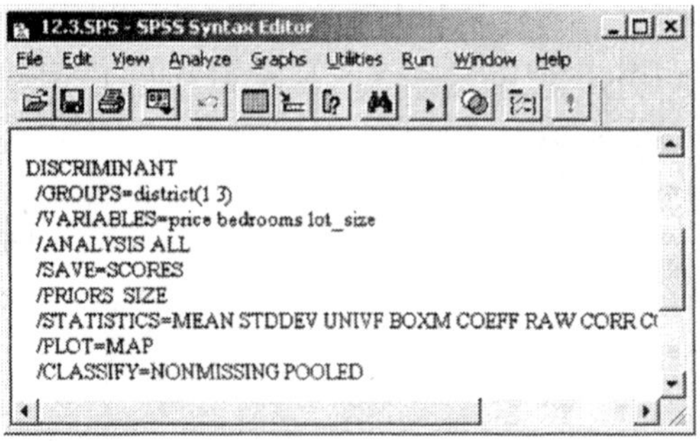

```
DISCRIMINANT
 /GROUPS=district(1 3)
 /VARIABLES=price bedrooms lot_size
 /ANALYSIS ALL
 /SAVE=SCORES
 /PRIORS SIZE
 /STATISTICS=MEAN STDDEV UNIVF BOXM COEFF RAW CORR C
 /PLOT=MAP
 /CLASSIFY=NONMISSING POOLED
```

Output 12.4.A *Classification results*

The Eigenvalues Table demonstrates that the first discriminant function (value 1.035) more effectively separates the data across groups than the second function (value .155). The Structure Matrix demonstrates that two variables, Lot size and Price, are most highly correlated with Function 1. Number of Bedrooms correlates most highly with the second discriminant function.

Eigenvalues

Function	Eigenvalue	% of Variance	Cumulative %	Canonical Correlation
1	1.035[a]	87.0	87.0	.713
2	.155[a]	13.0	100.0	.367

a. First 2 canonical discriminant functions were used in the analysis.

Structure Matrix

	Function	
	1	2
LOT_SIZE	.953*	-.206
PRICE	.483*	.041
BEDROOMS	.200	.839*

Pooled within-groups correlations between discriminating variables and standardized canonical discriminant functions
Variables ordered by absolute size of correlation within function.

*. Largest absolute correlation between each variable and any discriminant function

Standardized Coefficients allow some assessment of relative contribution of the variables to the functions. The unstandardized coefficients are not useful for comparison among variables.

Standardized Canonical Discriminant Function Coefficients

	Function	
	1	2
PRICE	.108	-.700
BEDROOMS	.227	1.250
LOT_SIZE	.947	.092

Canonical Discriminant Function Coefficients

	Function	
	1	2
PRICE	.001	-.006
BEDROOMS	.300	1.649
LOT_SIZE	.378	.037
(Constant)	-4.643	-3.036

Unstandardized coefficients

Sixty five percent of cases were correctly classified by this analysis.

Classification Results [a]

			Predicted Group Membership			
		DISTRICT	1	2	3	Total
Original	Count	1	7	1	1	9
		2	1	8	4	13
		3	1	4	8	13
	%	1	77.8	11.1	11.1	100.0
		2	7.7	61.5	30.8	100.0
		3	7.7	30.8	61.5	100.0

a. 65.7% of original grouped cases correctly classified.

Output 12.4.B *Plot of real estate data in discriminant space.*

Create the syntax to run a bivariate scatterplot of the discriminant scores (saved by the /SAVE = SCORES subcommand).

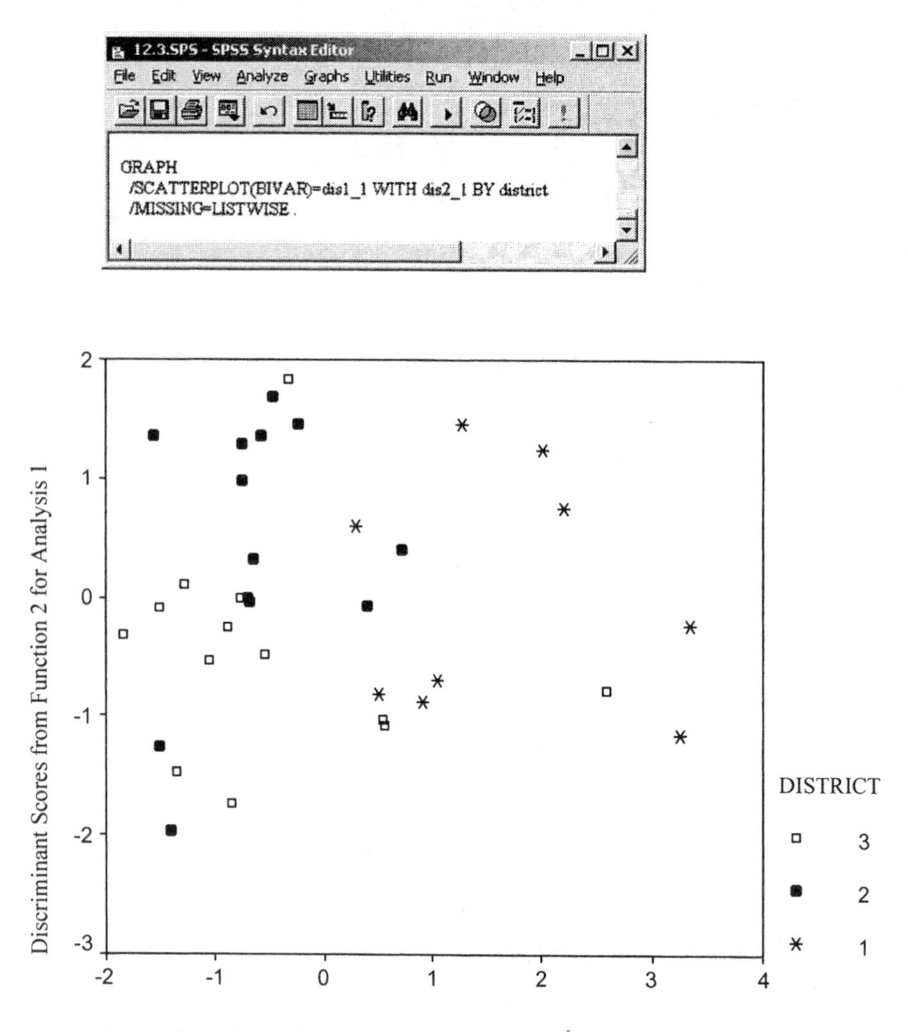

13 Logistic Regression Using SPSS

Example 13.1 *Logistic regression of likelihood to buy -- Books by Mail*

Logistic regression is applicable to similar research questions as discriminant analysis. Example 13.1 uses the Books by Mail data and can be compared to analyses presented in Chapter 12. In this example build a regression model with two predictors of book purchases.

Open the file Books_2.sav by double-clicking its icon. As in Example 12.1 rename variables col1 through col4 to custid, recent, artbooks and buy. Buy is the binary dependent variable, with values 0 and 1.

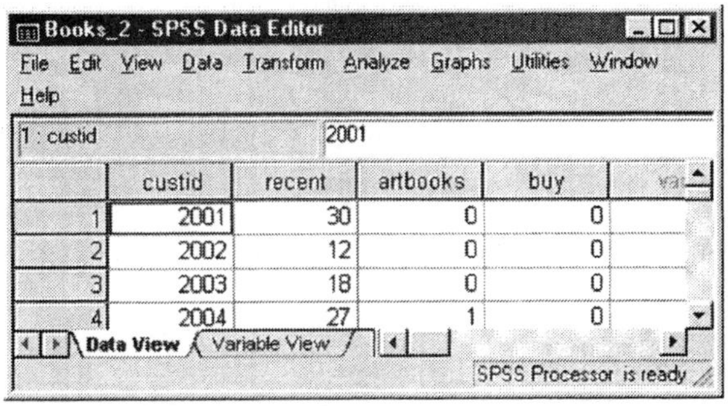

Books_2 - SPSS Data Editor

	custid	recent	artbooks	buy
1	2001	30	0	0
2	2002	12	0	0
3	2003	18	0	0
4	2004	27	1	0

In the **Analyze** menu select the **Regression-Binary Logistic** path.

In the **Logistic Regression** dialog box, select buy as the dichotomous dependent variable. Select recent and artbooks as covariates. Select from among several stepwise or block entry regression methods in the **Method** drop-down menu. For this exercise select the default method, Enter.

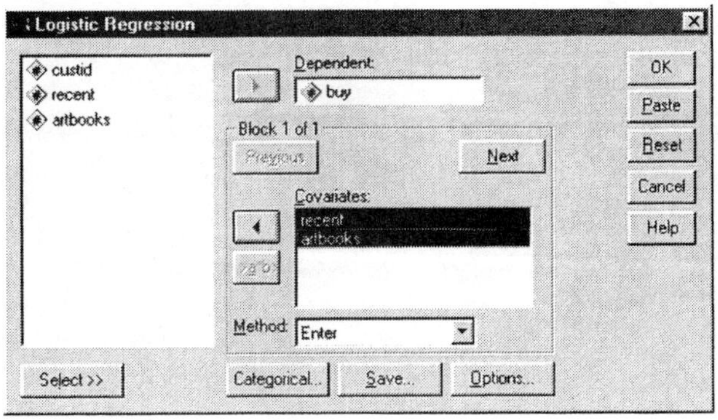

Click the **Options** button to specify classification plots and the Hosmer-Lemeshow Goodness of Fit statistic.

Other Logistic Regression Analysis Options

For categorical independent variables click the **Categorical** button to specify. Click the **Select>>**box to name a selection variable and specify selection rules for running a sub set of cases for analysis. For example, if the Books by Mail data set included population characteristics the logistic regression analysis might be run for female buyers, or customers between the ages of 35 and 45. The **Save** button permits saving predicted values, measures of influence and residuals into the working data file.

Pasted Syntax

```
*Open books_1.sav by double-clicking.
RENAME VARIABLES (col1 col2 col3 col4 = custid recent artbooks buy).

LOGISTIC REGRESSION VAR=buy
  /METHOD=ENTER recent artbooks
  /CLASSPLOT
  /PRINT=GOODFIT SUMMARY
  /CRITERIA PIN(.05) POUT(.10) ITERATE(20) CUT(.5) .
```

Output 13.1

Model summary terms are computed for each step in the logistic regression model. Each model summary term is a maximum likelihood-based approximation the r-square statistic, indicating how well the model explains variation. Like r-square terms each model summary term has a maximum of 1.

Model Summary

Step	-2 Log likelihood	Cox & Snell R Square	Nagelkerke R Square
1	471.656	.087	.202

The Hosmer and Lemeshow test is a goodness-of-fit indicator. Where significance is less than .05 the model does not adequately fit data. The Contingency Table for the Hosmer and Lemeshow test contains observed and predicted cases for each iteration of a model step. The contingency table contains the constituent terms used in computing the Hosmer and Lemeshow test statistic.

Hosmer and Lemeshow Test

Step	Chi-square	df	Sig.
1	11.343	8	.183

Contingency Table for Hosmer and Lemeshow Test

		BUY = 0		BUY = 1		
		Observed	Expected	Observed	Expected	Total
Step 1	1	94	96.049	3	.951	97
	2	115	113.321	1	2.679	116
	3	87	86.261	2	2.739	89
	4	93	92.579	3	3.421	96
	5	81	83.369	6	3.631	87
	6	104	102.505	4	5.495	108
	7	84	83.333	5	5.667	89
	8	88	91.786	12	8.214	100
	9	93	88.873	9	13.127	102
	10	80	80.923	36	35.077	116

The classification table cross tabulates observed and predicted outcomes. The model with SPSS default cut value of .5 predicts only 9.9% (11) of actual purchases.

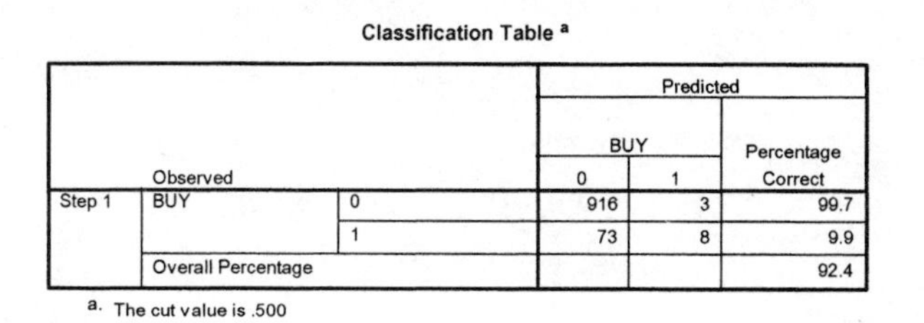

Classification Table [a]

	Observed		Predicted BUY 0	Predicted BUY 1	Percentage Correct
Step 1	BUY	0	916	3	99.7
		1	73	8	9.9
	Overall Percentage				92.4

a. The cut value is .500

A second model with cut value .160 does a better job at correctly predicting purchases (46.9% of purchases correctly predicted). A model with cut value .16 also incorrectly predicts a greater percentage of non-purchases (91). It correctly predicts 86.6% of results. Logistic regression and discriminant analysis classification tables can be compared for their predictive effectiveness (compare to Examples12.1 and 12.3).

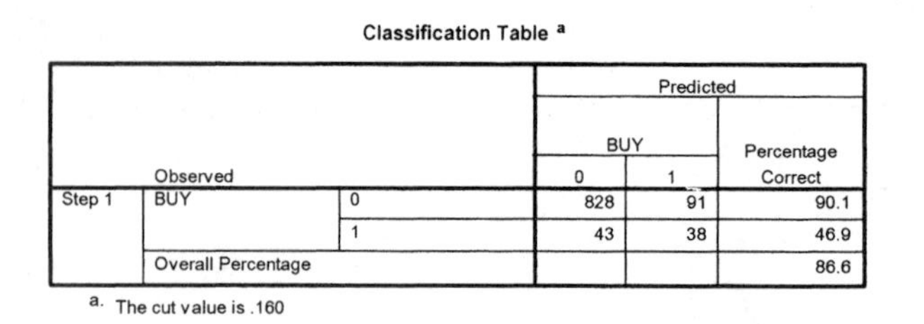

Classification Table [a]

	Observed		Predicted BUY 0	Predicted BUY 1	Percentage Correct
Step 1	BUY	0	828	91	90.1
		1	43	38	46.9
	Overall Percentage				86.6

a. The cut value is .160

The Variables in the Equation table summarizes the role of each model parameter, including estimated coefficients. The Wald statistic is the ratio of the Beta (B) over the Standard Error. Where the Wald is significant the parameter contributes to the model. Exp(B) is the predicted change in odds per unit change in a parameter.

Variables in the Equation

		B	S.E.	Wald	df	Sig.	Exp(B)
Step 1 [a]	RECENT	-.082	.019	17.898	1	.000	.921
	ARTBOOKS	1.214	.150	65.635	1	.000	3.368
	Constant	-2.275	.248	84.039	1	.000	.103

a. Variable(s) entered on step 1: RECENT, ARTBOOKS.

Example 13.2 *Stepwise Logistic Regression on ATM Adoption Data*

This example demonstrates stepwise logistic regression, where the model is progressively modified by introducing independent variables gradually into the model according to their score statistic ratings. This example will demonstrate forward stepwise selection method according to the Likelihood Ratio (LR) score statistic, based on maximum likelihood estimates for each parameter.

Open the file atm_adopt.sav and rename variable col2 to Adopt.

Select the **Analyze-Regression-Binary Logistic** path to open the **Logistic Regression** dialog box. Select adopt as the dependent variable. Select variables col3, col4, col5 as covariates. Select Forward LR from the **Method** drop-down menu.

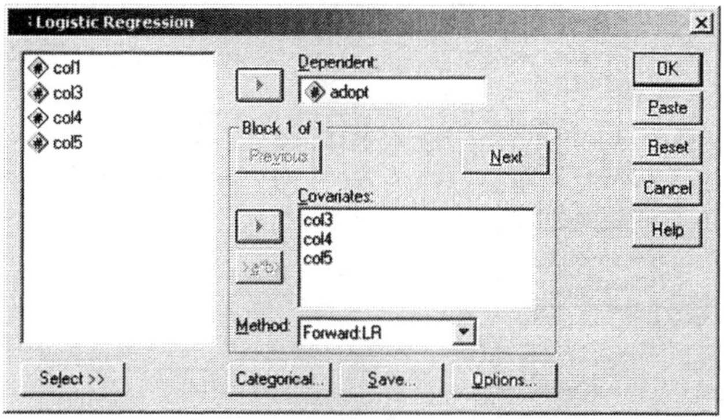

Logistic Regression
col1
col3
col4
col5
Dependent:
adopt
Block 1 of 1
Previous
Next
Covariates:
col3
col4
col5
Method: Forward:LR
OK
Paste
Reset
Cancel
Help
Select >>
Categorical...
Save...
Options...

Output 13.2

The Variables in the Equation Table summarizes the performance and contribution of parameters. Col5 is never significant, so it is not included in either of the model's two steps. The Iteration History table records the step criteria and model coefficients for each stage of the analysis. Iterations continue where the –2 log-likelihood ratio can be reduced. Step 1, in which the model is run with Col4 as independent variable, has four iterations. Once meaningful reductions in –2 log-likelihood ratio cease SPSS begins a second step, which includes both Col4 and Col3 as independent variables.

Variables in the Equation

		B	S.E.	Wald	df	Sig.	Exp(B)
Step 1 [a]	COL4	-1.735	.443	15.345	1	.000	.176
Step 2 [b]	COL3	.872	.330	6.981	1	.008	2.392
	COL4	-2.360	.516	20.952	1	.000	.094

a. Variable(s) entered on step1: COL4.

b. Variable(s) entered on step2: COL3.

Iteration History [a,b,c]

Iteration		-2 Log likelihood	Coefficients	
			COL4	COL3
Step 1	1	117.611	-1.400	
	2	117.000	-1.701	
	3	116.994	-1.734	
	4	116.994	-1.735	
Step 2	1	110.252	-1.882	.741
	2	109.324	-2.310	.864
	3	109.314	-2.359	.872
	4	109.314	-2.360	.872

a. Method: Forward Stepwise (Likelihood Ratio)

b. Initial -2 Log Likelihood: 138.629

c. Estimation terminated at iteration number 4 because parameter estimates changed by less than .001.

The Model if Term Removed table and Model summary statistics table summarize how much parameters contribute to the model. The Model if Term Removed table significance statistic indicates a parameter should be retained if significance is less than .05. The Model if Term Removed table contains Pseudo r-Square terms that indicate proportion of variation explained.

Model if Term Removed

Variable		Model Log Likelihood	Change in -2 Log Likelihood	df	Sig. of the Change
Step 1	COL4	.000	116.994	1	.000
Step 2	COL3	-58.497	7.680	1	.006
	COL4	-69.315	29.315	1	.000

Model Summary

Step	-2 Log likelihood	Cox & Snell R Square	Nagelkerke R Square
1	116.994	.195	.259
2	109.314	.254	.339

The Variables Not in the Equation table Overall Statistics row records tests at each step run to determine whether excluded parameter coefficients are equal to zero.

Variables not in the Equation

			Score	df	Sig.
Step 1	Variables	COL3	7.420	1	.006
		COL5	.097	1	.755
	Overall Statistics		9.855	2	.007
Step 2	Variables	COL5	2.803	1	.094
	Overall Statistics		2.803	1	.094

Example 13.3 Multinomial Logistic Regression

The NOMREG Multinomial Logistic Regression command offers a different set of analysis options from the LOGISTIC REGRESSION command. These include different goodness of fit tests, options for tests on sub populations of the data, different means to test parameters and capability to specify nested models. This example uses the ATM adoption data, as in Example 13.2.

Select the **Analyze-Regression-Multinomial Logistic** path to open the **Multinomial Logistic Regression** dialog box. Select adopt as the dependent variable. Select variables col3, col4, col5 as covariates. From the Statistics dialog box select Summary statistics, Likelihood ratio test, Parameter estimates, covariance and correlation matrices.

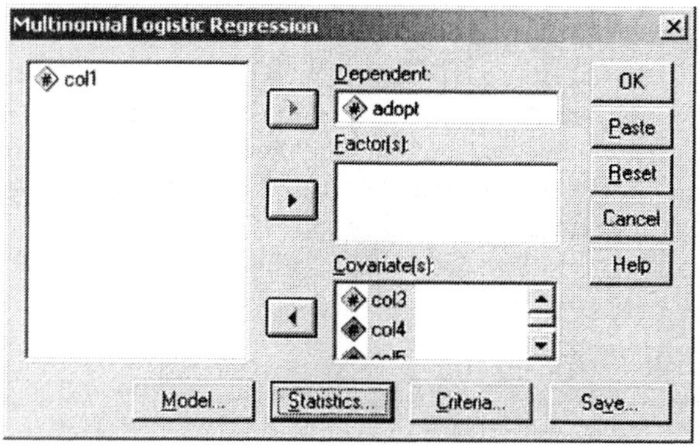

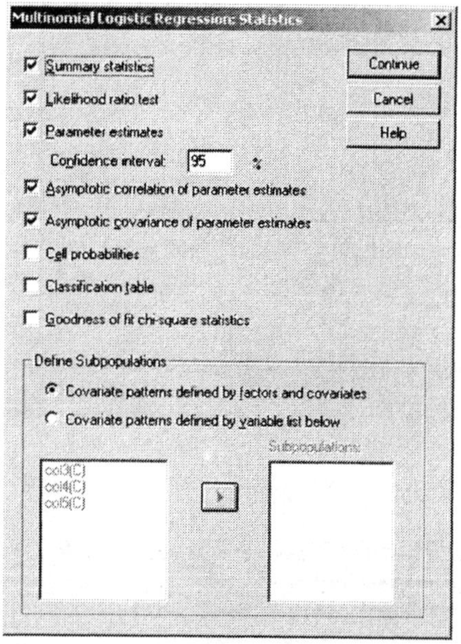

Pasted syntax

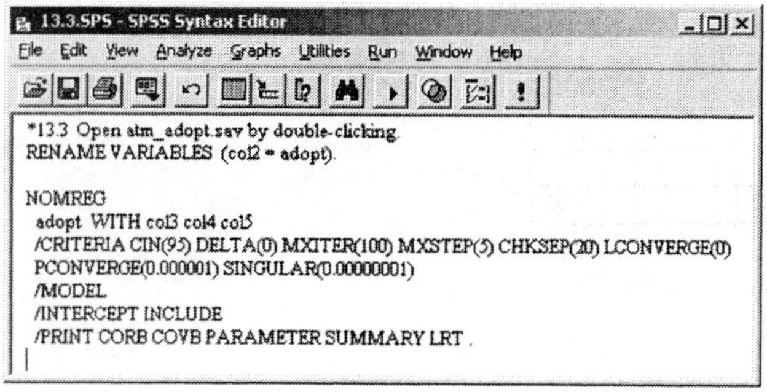

```
*13.3 Open atm_adopt.sav by double-clicking.
RENAME VARIABLES (col2 = adopt).

NOMREG
 adopt WITH col3 col4 col5
 /CRITERIA CIN(95) DELTA(0) MXITER(100) MXSTEP(5) CHKSEP(20) LCONVERGE(0)
 PCONVERGE(0.000001) SINGULAR(0.00000001)
 /MODEL
 /INTERCEPT INCLUDE
 /PRINT CORB COVB PARAMETER SUMMARY LRT .
```

Output 13.3

Model fitting information contains likelihood ratio tests against a null model (parameters = 0). The magnitude of Chi Square and degree of significance (less than .05) suggest whether the final model fits data better than the null model. The pseudo R-Square indicates the proportion of variation explained by the model.

Model Fitting Information

Model	-2 Log Likelihood	Chi-Square	df	Sig.
Intercept Only	55.748			
Final	24.228	31.521	3	.000

Pseudo R-Square

Cox and Snell	.270
Nagelkerke	.362
McFadden	.229

The Likelihood Ratio tests table indicates the relative contribution of each parameter to the model. The –2 Log Likelihood column figures are computed for the 'Reduced' model. For example, 28.186 is the –2 log-likelihood computed for the model with variable Col3 excluded. The Chi-square column numbers are reduced –2 log likelihood subtracted from final model –2 log likelihood. The Chi-square of 3.958 for Col3 suggests it contributes little to the final model. The Significance test measure is another indicator of contribution to final model. Col3 at .047 is just below the .05 mark.

Likelihood Ratio Tests

Effect	-2 Log Likelihood of Reduced Model	Chi-Square	df	Sig.
Intercept	24.654	.426	1	.514
COL3	28.186	3.958	1	.047
COL4	44.450	20.223	1	.000
COL5	27.327	3.099	1	.078

The chi-square statistic is the difference in -2 log-likelihoods between the final model and a reduced model. The reduced model is formed by omitting an effect from the final model. The null hypothesis is that all parameters of that effect are 0.

The Parameter Estimates table again, in the Significance column, indicates whether parameters contribute to the model. The positive or negative Beta coefficient sign suggests the likelihood of parameter unit changes in the model.

Parameter Estimates

ADOPT [a]		B	Std. Error	Wald	df	Sig.	Exp(B)	95% Confidence Interval for Exp(B) Lower Bound	Upper Bound
0	Intercept	-.281	.431	.425	1	.515			
	COL3	-1.069	.555	3.710	1	.054	.343	.116	1.019
	COL4	2.222	.543	16.748	1	.000	9.227	3.183	26.743
	COL5	.910	.522	3.042	1	.081	2.484	.893	6.908

a. The reference category is: 1.

Output includes correlation and covariance matrices of parameters.

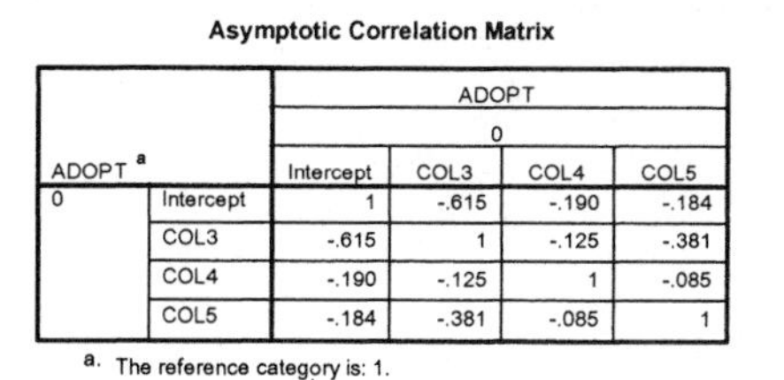

Asymptotic Correlation Matrix

ADOPT [a]		ADOPT 0 Intercept	COL3	COL4	COL5
0	Intercept	1	-.615	-.190	-.184
	COL3	-.615	1	-.125	-.381
	COL4	-.190	-.125	1	-.085
	COL5	-.184	-.381	-.085	1

a. The reference category is: 1.

Asymptotic Covariance Matrix [b]

ADOPT [a]		ADOPT 0 Intercept	COL3	COL4	COL5
0	Intercept	.186	-.147	-.044	-.041
	COL3	-.147	.308	-.038	-.110
	COL4	-.044	-.038	.295	-.024
	COL5	-.041	-.110	-.024	.272

a. The reference category is: 1.

b. There is no overdispersion adjustment.

The Observed and Predicted Frequencies table provides details of the model's predictive accuracy by combination of parameter

Observed and Predicted Frequencies

COL5	COL4	COL3	ADOPT	Frequency Observed	Predicted	Pearson Residual	Percentage Observed	Predicted
0	0	0	0	7	7.744	-.354	38.9%	43.0%
			1	11	10.256	.354	61.1%	57.0%
		1	0	6	4.117	1.041	30.0%	20.6%
			1	14	15.883	-1.041	70.0%	79.4%
	1	0	0	4	3.498	.758	100.0%	87.4%
			1	0	.502	-.758	.0%	12.6%
		1	0	4	5.641	-1.273	50.0%	70.5%
			1	4	2.359	1.273	50.0%	29.5%
1	0	0	0	2	1.305	1.033	100.0%	65.2%
			1	0	.695	-1.033	.0%	34.8%
		1	0	6	7.835	-.840	30.0%	39.2%
			1	14	12.165	.840	70.0%	60.8%
	1	0	0	9	9.454	-.631	90.0%	94.5%
			1	1	.546	.631	10.0%	5.5%
		1	0	17	15.407	1.069	94.4%	85.6%
			1	1	2.593	-1.069	5.6%	14.4%

The percentages are based on total observed frequencies in each subpopulation.